ECHNOLOGICAL CHANGE & LABOUR RELATIONS

Muneto Ozaki
et al.

TECHNOLOGICAL CHANGE & LABOUR RELATIONS

International Labour Office
Geneva

First published 1992

Ozaki, M., et al.
Technological change and labour relations
Geneva, International Labour Office, 1992
/Case study/s, /Comparative study/, /Technological change/, /Labour relations/, /Germany/ /Italy/, /Japan/, /Sweden/, /UK/, /USA/. 12.06.2
ISBN 92-2-107753-5

ILO Cataloguing in Publication Data

The designations employed in ILO publications, which are in conformity with United Nations practice, and the presentation of material therein do not imply the expression of any opinion whatsoever on the part of the International Labour Office concerning the legal status of any country, area or territory or of its authorities, or concerning the delimitation of its frontiers.
The responsibility for opinions expressed in signed articles, studies and other contributions rests solely with their authors, and publication does not constitute an endorsement by the International Labour Office of the opinions expressed in them.
Reference to names of firms and commercial products and processes does not imply their endorsement by the International Labour Office, and any failure to mention a particular firm, commercial product or process is not a sign of disapproval.

ILO publications can be obtained through major booksellers or ILO local offices in many countries, or direct from ILO Publications, International Labour Office, CH-1211 Geneva 22, Switzerland. A catalogue or list of new publications will be sent free of charge from the above address.

Printed in Switzerland

IDE

Preface

In 1987 and 1988 the ILO carried out a study on the interaction of technological change and labour relations consisting of the preparation of national monographs on six industrialised market economy countries, namely Germany,[1] Italy, Japan, Sweden, the United Kingdom and the United States. This book includes all six monographs together with a comparative paper written mainly, but not solely, on the basis of their findings.

Focusing on microelectronics technology, the study covers three industries, i.e. the machinery manufacturing, printing and banking industries. Each monograph consists of case-studies on selected enterprises in these industries. Although we initially intended to conduct, and publish, three case-studies from six countries, we have had to limit their number for reasons of space and because of the difficulty in obtaining access to suitable firms for investigation. Consequently, while the monographs on Japan and Sweden consist of case-studies from all three industries, those on Germany and the United States cover only the machinery manufacturing and the printing industries, and those on Italy and the United Kingdom the printing and the banking industries.

The study shows that throughout the introduction of technological change there was greater continuity and stability in labour relations sytems providing for problem-solving through workers' participation than in those relying exclusively, or predominantly, on the establishment and application of standards, for instance collective bargaining and grievance procedures. It also highlights the importance of flexible and broadly defined work organisation in ensuring that the former participative systems of labour relations continue to operate.

Grateful thanks are due to Caroline Hartnell, who ably edited and condensed the original manuscript.

Note

[1] Throughout this publication, "Germany" refers to the former Federal Republic of Germany.

Contents

Preface v

1. Technological change and labour relations: an international overview – *Muneto Ozaki* 1

1. Management and workers in the face of new technology 2

1.1 Managerial policies on new technology 2

1.1.1 Reduction of labour input 2
1.1.2 Improved efficiency 3
1.1.3 Improved quality of products and services 3
1.1.4 Customised production 4

1.2 Workers' response to new technology 5

1.2.1 Model technology agreements 5
1.2.2 Development of union policies on new technology 6

2. Labour-management interaction on the introduction of new technology 7

2.1 Methods of labour-management interaction 8

2.1.1 Methods of establishing procedural rules 8
2.1.2 Methods of solving problems 11

2.2 Impact of labour-management interaction on the planning of technological change 15

2.3 Impact of labour-management interaction on the consequences of introducing technological change 18

2.3.1 Job security 18
2.3.2 Job classification and income protection 21
2.3.3 Training and retraining 22

3. New technology and the reorganisation of work 25

3.1 Technology and work organisation 25

3.1.1 Teamwork and polyvalence 25
3.1.2 Skill structure 26
3.1.3 Working hours 27

3.2 Labour relations and work organisation 28

3.2.1 Methods and extent of workers' involvement 28
3.2.2 Influence of labour relations systems on decision-making 29

3.3 Influence of existing work organisation and skill structure 30
3.3.1 Work organisation and skill structure before technological change 30
3.3.2 Work organisation and skill structure after technological change 33

4. The influence of technological change on labour relations systems 34
4.1 Effects on the structure of the workforce 34
4.2 Effects on the unions 36
4.3 Effects on management 38
4.4 Patterns of negotiation and consultation 39

5. Concluding remarks 41

2. Technological change and labour relations in Germany – *Gert Schmidt* 47

1. Technological change and the labour relations system in Germany 47

2. Case-study of a printing firm 49
2.1 The context 49
2.1.1 The enterprise 49
2.1.2 Labour relations prior to the introduction of the new technology 50
2.1.3 General characteristics of employment relationships 51
2.1.4 Reasons for introducing the new technology 51
2.2 The decision-making process 52
2.2.1 Decision-making regarding the introduction of the new technology 52
2.2.2 The role of negotiation and consultation 52
2.3 Consequences for the workforce of introducing the new technology 53
2.3.1 Job security 53
2.3.2 Work organisation and working conditions 53
2.3.3 Payment systems and income protection 54
2.3.4 Training and retraining 54
2.4 Effects of the new technology on labour relations 55
2.4.1 Effects on the unions 55
2.4.2 Effects on management 55
2.4.3 Patterns of negotiation and consultation 55
2.4.4 Conflict 56
2.5 Evaluation 56

3. Case-study of a machine-tool manufacturer 57
3.1 The context 57
3.1.1 The enterprise 57
3.1.2 Labour relations prior to the introduction of the new technology 58

3.1.3 General characteristics of employment relationships 60
3.1.4 Reasons for introducing the new technology 60
3.2 The decision-making process 61
3.2.1 Decision-making regarding the introduction of the new technology 61
3.2.2 The role of negotiation and consultation 62
3.3 Consequences for the workforce of introducing the new technology 63
3.3.1 Job security and income protection 63
3.3.2 Work organisation and working conditions 64
3.3.3 Training and retraining 64
3.4 Effects of the new technology on labour relations 64
3.4.1 Effects on the structure of the workforce 64
3.4.2 Effects on the unions 65
3.4.3 Effects on management 65
3.4.4 Patterns of negotiation and consultation 65
3.4.5 Conflict 65
3.5 Evaluation 66

3. Technological change and labour relations in Italy 69

1. Case-study of a national newspaper – *Aldo Marchetti* **69**
1.1 The context 69
1.1.1 The newspaper industry 69
1.1.2 The enterprise 70
1.1.3 Labour relations prior to the introduction of the new technology 70
1.1.4 Reasons for introducing the new technology 70
1.2 The decision-making process 71
1.2.1 Decision-making regarding the introduction of the new technology 71
1.2.2 The role of negotiation and consultation 72
1.3 Consequences for the workforce of introducing the new technology 73
1.3.1 Job security 73
1.3.2 Work organisation and working conditions 74
1.3.3 Payment systems and income protection 74
1.3.4 Training and retraining 75
1.4 Effects of the new technology on labour relations 75
1.4.1 Effects on the structure of the workforce 75
1.4.2 Effects on the unions 76
1.4.3 Effects on management 76
1.4.4 Patterns of negotiation and consultation 76
1.4.5 Conflict 76
1.5 Evaluation 77

2. Case-study of a bank – ***Tiziano Treu*** **77**

2.1 The context 77

2.1.1 The banking industry 77
2.1.2 The enterprise 78
2.1.3 Labour relations prior to the introduction of the new technology 78
2.1.4 General characteristics of employment relationships 79

2.2 The decision-making process 80

2.2.1 Decision-making regarding the introduction of the new technology 80
2.2.2 The role of negotiation and consultation 81

2.3 Consequences for the workforce of introducing the new technology 83

2.3.1 Job security 83
2.3.2 Work organisation and working conditions 83
2.3.3 Payment systems and income protection 84
2.3.4 Training and retraining 85

2.4 Effects of the new technology on labour relations 86

2.4.1 Effects on the structure of the workforce 86
2.4.2 Effects on the unions 86
2.4.3 Effects on management 86
2.4.4 Patterns of negotiation and consultation 87
2.4.5 Conflict 87

2.5 Evaluation 87

4. Technological change and labour relations in Japan – ***Yasuo Kuwahara*** **89**

1. Introduction **89**

2. Case-study of a printing firm **90**

2.1 The context 90

2.1.1 The printing industry 90
2.1.2 The enterprise 91
2.1.3 Labour relations prior to the introduction of the new technology 91
2.1.4 Reasons for introducing the new technology 92

2.2 The decision-making process 92

2.2.1 Decision-making regarding the introduction of the new technology 92
2.2.2 The role of negotiation and consultation 93

2.3 Consequences for the workforce of introducing the new technology 94

2.3.1 Job security 94
2.3.2 Work organisation and working conditions 94

2.3.3 Payment systems 95
2.3.4 Training and retraining 95

2.4 Effects of the new technology on labour relations 96

2.4.1 Effects on the structure of the workforce 96
2.4.2 Effects on the unions 96
2.4.3 Effects on management 96
2.4.4 Patterns of negotiation and consultation 97
2.4.5 Conflict 97

3. Case-study of a bank **97**

3.1 The context 97

3.1.1 The banking industry 97
3.1.2 Labour relations in the banking industry 99
3.1.3 The enterprise 99
3.1.4 General characteristics of employment relationships 99

3.2 The decision-making process 100

3.2.1 Decision-making regarding the introduction of the new technology 100
3.2.2 The role of negotiation and consultation 100

3.3 Consequences for the workforce of introducing the new technology 101

3.3.1 Job security 101
3.3.2 Work organisation and working conditions 102
3.3.3 Training and retraining 102

3.4 Effects of the new technology on labour relations 102

3.4.1 Effects on the structure of the workforce 102
3.4.2 Effects on the unions 103
3.4.3 Effects on management 103
3.4.4 Patterns of negotiation and consultation 103

4. Case-study of a machine-tool manufacturer **104**

4.1 The context 104

4.1.1 The machine-tool industry 104
4.1.2 The enterprise 105
4.1.3 Labour relations prior to the introduction of the new technology 105

4.2 The decision-making process 105

4.2.1 Decision-making regarding the introduction of the new technology 105
4.2.2 The role of negotiation and consultation 106

4.3 Consequences for the workforce of introducing the new technology 107

4.3.1 Job security 107
4.3.2 Work organisation 107
4.3.3 Payment systems and income protection 108
4.3.4 Training and retraining 108

4.4 Effects of the new technology on labour relations 109

4.4.1 Effects on the structure of the workforce 109
4.4.2 Effects on the unions 109
4.4.3 Patterns of negotiation and consultation 109

5. Evaluation **110**

5. Technological change and labour relations in Sweden – ***Bernd Hofmaier in collaboration with Kristina Hakansson*** **113**

1. Labour relations in Sweden **113**

1.1 Labour laws and agreements 113

1.2 Swedish unions and technological change 114

2. Case-study of a national newspaper **115**

2.1 The context 115

2.1.1 The newspaper industry 115
2.1.2 Labour relations in the newspaper industry 116
2.1.3 The enterprise 117
2.1.4 Labour relations prior to the introduction of the new technology 118
2.1.5 Reasons for introducing the new technology 118

2.2 The process 118

2.2.1 The initiative 118
2.2.2 The role of the project groups 119

2.3 Consequences for the workforce of introducing the new technology 120

2.3.1 Job security 120
2.3.2 Work organisation and working conditions 120
2.3.3 Payment systems 121
2.3.4 Training and retraining 121

2.4 Effects of the new technology on labour relations 122

3. Case-study of a bank **122**

3.1 The context 122

3.1.1 The banking industry 122
3.1.2 Labour relations in the banking industry 123
3.1.3 The enterprise 124
3.1.4 Labour relations prior to the introduction of the new technology 124
3.1.5 Reasons for introducing the new technology 125

3.2 The process 125

3.2.1 The initiative 125
3.2.2 The role of the project group 125

3.3 Consequences for the workforce of introducing the new technology 126

3.3.1 Job security 126
3.3.2 Work organisation 126
3.3.3 Payment systems 126

3.3.4 Training and retraining 126
3.4 Effects of the new technology on labour relations 126

4. Case-study of a machinery manufacturer **127**
4.1 The context 127
4.1.1 The enterprise 127
4.1.2 The development project at D3 128
4.1.3 Labour relations prior to the start of the project 128
4.2 The process 129
4.2.1 The steering committee 129
4.2.2 The project group 129
4.2.3 The development groups 129
4.2.4 Setting social, technical and economic targets 130
4.2.5 The development group reports 130
4.2.6 Results so far 131
4.3 Consequences for the workforce of introducing the new technology 132
4.3.1 Job security 132
4.3.2 Work organisation 132
4.3.3 Payment systems 133
4.3.4 Training and retraining 133
4.4 Effects of the new technology on labour relations 133

5. Evaluation **134**

6. Technological change and labour relations in the United Kingdom – *Roderick Martin in collaboration with Michael Noon* **137**

1. Introduction – *Roderick Martin* **137**
1.1 National provisions applying to labour-management relations with respect to the introduction of new technology 137
1.2 Impact of microelectronics on the workforce 138

2. Case-study of a regional newspaper – *Michael Noon* **139**
2.1 The context 139
2.1.1 The enterprise 139
2.1.2 Labour relations prior to the introduction of the new technology 140
2.2 The decision-making process 141
2.2.1 Decision-making regarding the introduction of the new technology 141
2.2.2 The role of negotiation and consultation 142
2.3 Consequences for the workforce of introducing the new technology 146
2.3.1 Job security 146
2.3.2 Work organisation and working conditions 147
2.3.3 Payment systems and income protection 148

2.3.4 Training and retraining 149

2.4 Effects of the new technology on labour relations 149

2.4.1 Effects on the structure of the workforce 149
2.4.2 Effects on the unions 149
2.4.3 Effects on management 151
2.4.4 Patterns of negotiation and consultation 151
2.4.5 Conflict 152

2.5 Evaluation 152

3. Case-study of a major national clearing bank – *Roderick Martin* **153**

3.1 The context 153

3.1.1 The enterprise 153
3.1.2 Reasons for introducing the new technology 153
3.1.3 Labour relations prior to the introduction of the new technology 155
3.1.4 General characteristics of employment relationships 157

3.2 The decision-making process 158

3.2.1 Decision-making regarding the introduction of the new technology 158
3.2.2 The role of negotiation and consultation 158

3.3. Consequences for the workforce of introducing the new technology 160

3.3.1 Job security 160
3.3.2 Work organisation and working conditions 160
3.3.3 Payment systems and income protection 161
3.3.4 Training and retraining 162

3.4 Effects of the new technology on labour relations 162

3.4.1 Effects on the structure of the workforce 162
3.4.2 Effects on the unions 163
3.4.3 Effects on management 164
3.4.4 Patterns of negotiation and consultation 164
3.4.5 Conflict 165

3.5 Evaluation 166

7. Technological change and labour relations in the United States – *Daniel B. Cornfield* **169**

1. Case-study of newspaper composing rooms in Chicago **169**

1.1 The context 170

1.1.1 Technological and labour relations developments in newspaper composing rooms 170
1.1.2 Labour relations in newspaper composing rooms in Chicago 172

1.2 The process 173

1.2.1 Decision-making regarding the introduction of the new technology 173
1.2.2 Negotiations over control of the workplace 174

1.3 Consequences for the workforce of introducing the new technology 177
1.3.1 Job security 177
1.3.2 Work organisation 179
1.3.3 Payment systems 179
1.3.4 Training and retraining 180
1.4 Evaluation 182

2. Case-study of a die manufacturer **183**
2.1 Technological and labour relations developments in the metalworking machinery manufacturing industry 183
2.1.1 The diffusion of numerical control 184
2.1.2 Employment trends in metalworking machinery manufacturing 184
2.1.3 Occupational employment, skill and earnings 187
2.1.4 Developments in labour relations 189
2.1.5 Summary 192
2.2 Context for the introduction of numerical control at a die manufacturer 193
2.2.1 The enterprise 193
2.2.2 Labour relations prior to the introduction of the new technology 194
2.2.3 General characteristics of employment relationships 194
2.3 The decision-making process 195
2.3.1 Decision-making regarding the introduction of the new technology 195
2.3.2 The role of negotiation and consultation 195
2.4 Consequences for the workforce of introducing the new technology 197
2.4.1 Job security 197
2.4.2 Work organisation 197
2.4.3 Payment systems and income protection 198
2.4.4 Training and retraining 199
2.5 Effects of the new technology on the structure of the workforce 199
2.6 Evaluation 201

1

Technological change and labour relations: an international overview

*Muneto Ozaki**

This introductory chapter, like the national monographs in the following six chapters, analyses the reciprocal influence that technological change and labour relations exert on each other. In more concrete terms, it describes the extent to which and the ways in which workers or their representatives participate in managerial decision-making concerning the introduction of new technology. It also evaluates the degree of effectiveness of such participation, and analyses the ways in which new technology in turn affects the extent, forms and effectiveness of workers' participation.

The social impact of technological change, for example its impact on employment, skill levels and workforce structure, is dealt with in this chapter only in so far as this is necessary for the analysis of the extent to which it can be modified by labour relations, or only in so far as it in turn affects the patterns of labour relations. Although many studies already exist on the social impact of technological change, there seems to be a relative scarcity of studies on the reciprocal influence of labour relations and technological change. This study attempts to fill this gap.

The term "new technology", as used in this book, calls for an explanation. This study focuses on a particular form of new technology, that is, microelectronics technology, because of its pervasiveness in industry today and the extent of its impact on workers. The case-studies on the metalworking industry mainly analyse the changes related to the introduction of numerically controlled (NC) machine tools, computerised numerically controlled (CNC) machine tools, industrial robots and flexible manufacturing systems (FMS). The case-studies on the printing industry primarily analyse the change from hot-metal typesetting to photocomposition and the introduction of computerised typesetting, ultimately leading to the elimination of the boundaries between editorial and typesetting work; in the area of colour and picture printing, they mainly study the introduction of computerised colour scanning, which enables colours to be separated into points with varying numerical values and recomposed through computers. The case-studies on the banking industry focus on the installation of on-line terminals at the counter and the use of "back-office" automation; they deal only peripherally with other applications of computers such as automated telling machines (ATMs), electronic funds transfer at the point of sale (EFT/POS), and home banking.

The first section of this chapter analyses the policies of management and unions on new technology. The second section deals with the methods and the extent of the

*International Labour Office, Geneva.

involvement of workers and unions in decision-making concerning technological change. The third section analyses the ways in which work processes are reorganised after (or at the time of) the introduction of technological change, under the influence of a variety of factors including labour relations. The fourth section deals with the influence technological change has on labour relations systems.

1. Management and workers in the face of new technology

1.1 Managerial policies on new technology

In market economies, the ultimate goal of corporate managerial strategies is the creation of profits, which is the basic condition for the survival of the enterprise. Management invest in new technology to attain this goal, hoping that new technology will give the enterprise a competitive edge over its rivals in the market. However, the concrete objectives sought by management through specific technological changes vary widely. The available data nevertheless suggest that, through technological change, management usually seek to attain one or several of the following objectives:

(1) a reduction of labour input in work processes, either in order to reduce labour costs or to cope with labour shortages;
(2) greater efficiency of operations through closer managerial control over production processes;
(3) higher quality of products or services through the greater precision of operations and speedier delivery of information that the computer makes possible;
(4) improvement of the ability to produce custom-built products in small batches, and to adapt production to the diverse and changing demands of clients.

1.1.1 Reduction of labour input

Computerised machines may replace human labour on a large scale. Some engineers have even claimed that the ultimate purpose of computerised manufacturing systems is "wrestling manufacture away from human interference".[1] Although most employers today would not share such a radical view, many have in mind what is commonly referred to as the "labour-saving effect" of microelectronics technology when they introduce it into their workplace.

In some cases, for example, generally in Japan in the early 1970s and particularly in the German and Japanese printing firms and the German machine-tool manufacturer studied in later chapters in this book, management rely on the "labour-saving effect" of microelectronics technology to overcome labour shortages. More often, however, new technology is introduced to reduce labour costs. In the banking industry, for example, the massive computerisation of banking operations in recent years has been prompted by management's efforts to contain labour costs in the context of a steady rise in the level of banking transactions and growing inter-bank competition for customers. In the newspaper industry, the stagnating market and the rising cost of raw materials have incited publishers to introduce new technology with a view to cutting production costs through the elimination of human labour from production processes, in particular in the composing room. The reduction of labour costs is a pervasive objective of technological change in the metalworking industry also.

1.1.2 Improved efficiency

Microelectronics technology may also be instrumental in ensuring effective managerial control over the whole range of production processes, and thus contribute to the enhancement of managerial efficiency. The use of computers for stock control is today widespread.

A more controversial issue is the potential use of new technology by employers as a means of breaking down traditional skills and degrading human labour with a view to tightening their control over work processes. Whether or not employers actually seek these objectives when they introduce new technology depends on a number of factors and varies widely from one country or enterprise to another.

In general, where employers pursue the Taylorist goals of fragmenting and tightly controlling work, microelectronics technology may be used as an instrument for making many traditional skills redundant and giving management a virtual monopoly over new skills and knowledge, which facilitates their control of each step of the labour process and its mode of execution. This was predominantly the case in the early days of the massive computerisation of industrial work in many countries. Very often, such policies involved attempts to degrade work. During the late 1940s and early 1950s, for example, the management community in the United States is reported to have embraced numerical control more readily than record-playback (where the movements of a skilled and experienced worker are recorded on to a machine, and a previously manual task is thus converted into an automated mechanical one) because it led to greater reductions in the discretion exercised by skilled machinists and in management dependence on them.[2]

Things have been changing significantly recently, however, employers becoming increasingly conscious of the need to have skilled workers in order to benefit fully from new technology. Nevertheless, a number of craft skills (for example, those of typographers) and other traditional skills (for example, those of dockers), which used to give their holders a certain discretion in organising their work, have mostly been eliminated by the computerisation of work, and the new skills needed for operating computerised machines are increasingly under management control. There seems to have been a transfer of the sources of skills and knowledge from workers to management.

1.1.3 Improved quality of products and services

With computerised machines, it is possible to work with greater precision and at higher speed than with conventional machines. Therefore, management often seek to improve the quality of the products the enterprise produces or of the services it provides through the introduction of microelectronics technology. The German and United States machine-tool case-studies in later chapters show that this was one of the main objectives of the changes introduced into the enterprises studied. The Japanese and German printing case-studies also show that one aim of introducing computerised colour scanning into the picture printing process has been to improve the quality of picture printing. In the banking industry, inter-bank competition for customers has been intense, and banks are seeking to improve their advisory services to customers by introducing on-line systems to link the terminals at the counter to the computerised customer files in the central office, thus enabling staff to decide quickly, in the light of information on the customer, how to advise them (for example, on loans).

1.1.4 Customised production

The growing inter-firm competition in product markets makes it necessary for each enterprise to be able to deliver products that suit the particular requirements of each customer. Microelectronics technology is highly attractive from this point of view because a single computerised production facility can carry out a range of operations and change their specification with the minimum of cost and delay. It thus enables the producer both to produce custom-built products in small batches, and to respond quickly to changes in market demand as regards both models and quantities.

The importance of this objective is borne out by a number of case-studies. It was particularly prominent, for example, in the introduction of a flexible manufacturing system (FMS) into a German establishment producing transmission equipment for cars,[3] and in the massive introduction of industrial robots in the second half of the 1970s into the Japanese car industry (including parts producers).[4] In the newspaper industry, new technology has enabled national papers to develop regional editions at relatively low costs.

1.2 Workers' response to new technology

In the early days of the massive introduction of microelectronics technology into industry, union policies on new technology were basically defensive in that they focused on the consequences of technological change, unions seeking to protect workers against the adverse effects of new technology, in particular against possible job losses.

Since the middle of the 1970s, however, there have been significant developments in union policies on new technology in a number of countries. As competition becomes increasingly harsh, a growing number of workers and unions have come to realise that technological innovation is not only inevitable but may be beneficial to workers because it may improve the competitiveness of the enterprise, and consequently enhance job security and improve employment conditions. In a survey of 2,000 workplaces, conducted in 1984 in the United Kingdom, the manual workers directly affected by the introduction of microelectronics technology were reportedly in favour of the change in 74 per cent of cases, while the office workers were in favour of the change in 78 per cent of cases.[5]

Consequently, unions are today increasingly seeking to influence the process of technological innovation so that new technology may be introduced in such a way as to benefit workers and minimise its adverse consequences. In some countries, union central organisations have attempted to do so by drafting model agreements, and encouraging their affiliates to include them in their collective agreements. In others, unions have sought to develop their own technology policies, and attempted to participate in the process of technological innovation so as to influence the changes in line with their policies.

1.2.1 Model technology agreements

The efforts of unions to influence the process of technological change by means of collective bargaining have led in many countries to the conclusion of what are commonly called technology agreements or to the insertion of technology-related clauses in general collective agreements. This issue will be discussed later in this chapter, in the context of the examination of the role of labour relations in technological change. In some countries, however, mainly the Anglo-Saxon countries with a tradition of adversarial collective bargaining, union central organisations have formulated models for technology agreements, and encouraged their affiliate unions to include them in their collective agreements.

In the United Kingdom, for example, the 1979 Trades Union Congress (TUC) adopted a check-list for new technology agreements, which recommends that TUC affiliates incorporate in technology agreements such items as prior consultation, technological change by agreement, disclosure of information, guarantees of job security, expansion of output after technological change, commitments on retraining, reductions in working hours, improvements in employment conditions, health and safety standards, and procedures for reviewing progress. In addition, the check-list suggests that negotiations should seek to influence the design of new equipment.

In the United States, as our American machine-tool case-study reports, the International Association of Machinists (IAM) presented at its 1981 Electronics and New Technology Conference a "model contract language" (that is, model clauses for collective agreements), encouraging its local unions to bargain for their inclusion in collective agreements. It provides for advance notice, the establishment of a joint union-management committee for technological change, workforce reduction by attrition rather than lay-off when such a reduction is unavoidable, a commitment to retraining, seniority-based inter-establishment transfer rights, prevention of the classification of new jobs as non-bargaining unit jobs, and the automatic recognition of the IAM in new plants. The IAM also published a Workers' Technology Bill of Rights in 1981, setting out the principles that should govern the introduction of new technology.

In Japan, several union federations, such as the All Japan Federation of Electric Machine Workers' Unions (Denki-Roren) in 1984 and the National Federation of Metal Industry Trade Unions (Zenkin-Domei) in 1982, have drawn up guidelines for negotiators on the introduction of new technology.

Most of the clauses in the TUC check-list, the IAM model contract language and the Japanese guidelines are confined to dealing with the consequences of technological change, and did not constitute any major innovation in union bargaining strategy. However, the suggestion in the TUC check-list that negotiations should seek to influence the design of new equipment, and the provisions on advance notice and joint consultation in the IAM model contract language, were new to the labour relations systems of the respective countries.

As shown later in this chapter, the effects of these model technology agreements on the practice of collective bargaining seem to have been very limited.

1.2.2 Development of union policies on new technology

In some of the countries covered by this project, unions have made significant efforts to develop coherent policies on technology, aimed at making new technology instrumental in humanising work and generally enhancing workers' well-being as well as in improving the efficiency of industry. The basic assumption underlying such union strategies is that there is no scientific law that determines the direction of technological development, so it should be possible to develop technology that is beneficial socially as well as economically. According to the unions in the former Federal Republic of Germany (hereafter referred to as Germany), their role in technological change is, therefore, to ensure that technology remains "compatible with society" (sozialverträglich).[6]

In Sweden and Germany, where union efforts in this area are most notable, the new union policies on technology emerged in the first half of the 1970s. In Sweden, after the collapse of a major joint effort at creating new forms of work organisation (the URAF programme), unions concentrated their efforts on legislative reforms aimed at creating the framework for enabling them to participate in decision-making on technological change. This led to the passing in 1976 of the Co-determination Act (MBL). Under this Act, as discussed in more detail later in this chapter, unions in a number of enterprises have been participating in joint project groups for developing new technology and new forms of work organisation.

In Germany, new union policies have involved a shift of emphasis in union bargaining policies from quantitative objectives (related to the level of pay) towards qualitative objectives (aimed at infuencing the content of human labour). In more concrete terms, unions today demand that new technology should be introduced in such a way as to make either of the following two types of work organisation possible: first, integrated jobs (*Mischarbeit*), in which planning, preparation, execution and supervision are integrated into one job; second, group work, in which broadly and highly skilled homogeneous workers enjoy a high degree of discretion in distributing and rotating work among themselves. Their emergence was roughly concomitant with the launching in 1974 of the Humanisation of Working Life Programme under a centre-left coalition government, with the participation of the German Confederation of Trade Unions (Deutsche Gewerkschaftsbund – DGB),[7] although it also closely followed the emergence of new organisational philosophies among German employers, which favoured the development of flexible work organisation. Somewhat similar developments have been taking place in other Scandinavian trade unions also.[8]

The pursuit of these new policies makes it necessary for unions to build up technical resources within unions and among workers. Swedish unions seem to have achieved significant results in this area through group work among union members. A prominent example is the experience of union members in a new Scanian Dairies (Skanemejerier) dairy in Malmö, which started operating in April 1984.[9] Before the management's own planning work for the new dairy got under way seriously in 1980, the employees prepared themselves through union group work, formulating their goals and demands, analysing managerial strategies, and identifying the objectives that could realistically be attained. Such work was facilitated by a research project in which outside researchers as well as central union organisations took part. Through these actions, and also by gaining the right to be represented in the management body (project management group) responsible for making decisions on technological and organisational changes, the unions exerted a very important influence on the changes that were introduced into the Malmö dairy.

In Germany, the new union policies have created an acute need for the training of works council members, who are entrusted with the task of exercising some control over the actual organisation of work and technology within each enterprise in the light of the standards established by collective agreements, which is quite a new task for most works council members. In order to help them, German unions are producing models of "humane" work and training works council members in the proper use of the existing legal instruments for influencing work organisation. They have also established advisory centres on technology, where experts help works councils in examining management plans for changes and eventually formulating alternative proposals. The centres are funded by the Humanisation of Working Life Programme, mentioned earlier.

In other countries, the development of union policies on technological development has been less conspicuous. In many cases, unions are reluctant to get involved in managerial policy-making, and prefer to concentrate on dealing with the consequences of changes.

Nevertheless, the economic recession of the 1980s and the intensifying international competition have brought about a new trade union culture that is encouraging some trade unions to assume greater responsibility as regards technological innovation. In Italy such a change in union culture is illustrated by the signing in December 1984 of a protocol between the Institute of Industrial Reconstruction (Istituto Ricostruzione Industriale, IRI – the largest holding company for public sector enterprises) and the three major union confederations, which provides for the involvement of the unions in most decisions on reorganisation and technological innovation from the planning phase. In France, the French Democratic Confederation of Labour (Confédération Française Démocratique du Travail – CFDT) has made significant efforts to develop strategies for influencing the development and introduction of new technology.[10]

2. Labour-management interaction on the introduction of new technology

This section analyses (1) the methods of labour-management interaction on technological change, (2) the extent of workers' and unions' influence on the planning of technological change, and (3) the impact of labour-management interaction on the consequences of technological change, such as employment protection, income protection and retraining. The impact of labour-management interaction on work organisation will be discussed in the next section, as part of the analysis of the trilateral causal relationship between technological change, work organisation and labour relations.

2.1 Methods of labour-management interaction

When we discuss the methods of labour-management interaction on technological innovation, we need to make a distinction between labour-management interaction in order to establish procedural rules for the introduction of technological change and labour-management interaction in order to solve problems arising in the process of introducing such changes.

2.1.1 Methods of establishing procedural rules

The normal sources of procedural rules are legislation and collective agreements. The respective role of these two types of instrument varies from one country to another. In such countries as Sweden, Germany and France, legislation has established the basic right of workers to participate in technological changes, and collective agreements have improved upon it. In other countries, for example, Italy, the United Kingdom, the United States and Japan, collective agreements have been the main source of procedural rules for technological innovation.

Legislation and collective bargaining

Within the group of countries where legislation has established the basic right of workers to participate in technological change, the respective role of legislation and collective agreements in the establishment of procedural rules varies. In Sweden, the basic rules were established by the Co-determination Act of 1976, but collective bargaining has played a very important role as well. The Act established the employers' "primary duty of negotiation", that is, an obligation to take the initiative in entering into negotiations with the union (with which they have a collective agreement) on any plan to introduce major changes into the enterprise, such as new working methods or organisational or technological changes. Under this Act, a central co-determination agreement for the private sector called the Efficiency and Participation Agreement (UVA), was signed in 1982. It provides the framework for labour-management cooperation within enterprises in developing high efficiency, secure employment, a good working environment, stimulating work, and equality between men and women. In a growing number of enterprises, parties have been concluding local co-determination agreements in order to translate the spirit of the central agreement into concrete arrangements for labour-management cooperation, which normally takes the form of joint union-management study projects.

The spread of local co-determination agreements has been uneven; it has been quick in large workplaces with strong local trade unions, but slow in small workplaces. However, labour-management cooperation seems to be extensively practised informally even in places where there is no written agreement for this purpose.[11]

It is also noteworthy that, although labour-management cooperation through joint projects has gained impetus under the Co-determination Act and UVA, agreements providing for the establishment of joint project groups on technology issues already existed prior to the conclusion of UVA. The technical agreement of 1974 in the newspaper industry, to which our case-study refers, is an example.

In Germany, also, the basic rules for labour-management interaction on the introduction of technological change are provided by legislation, but, in a number of industries, collective bargaining has also played a significant role in shaping the rules. The workers' right to participate in technological change is based on the Works Constitution Act of 1972, which makes it incumbent upon employers to inform the works council in full and in good time of any alterations that may entail substantial prejudice to the employees, and consult it on the planned alterations. This consultative right of the works council as regards technological change has been clarified or strengthened by a new type of collective agreement that has emerged since the late 1960s, namely the "rationalisation protection agreement", which today covers a number of major industries such as the banking, printing, metal and chemical industries as well as the public service.

While most rationalisation protection agreements in the major industries are confined, in so far as the consultative rights of the works council on technological change are concerned, to clarifying the provisions of the Works Constitution Act, some others (for example, those in the cigarette industry and the North Rhine Westphalian Regional Development Corporation) have considerably strengthened the consultative rights by providing for binding arbitration of disagreements between the parties.[12] The Technology Agreement of 1987 between Volkswagen AG and the Metal Workers' Union (IG Metall) also provides for mediation of unresolved disagreements between management and works council.[13]

In France an Act of 1982 explicitly recognised the right of the works council (comité d'entreprise) to be informed and consulted on any major plans for introducing new technology. Collective agreements have also come to deal with technological change. A national inter-occupational agreement on new technology was concluded in 1988, while a few industry-level agreements were concluded earlier, for example, in banking in 1986 and in the metal industry in 1987. An agreement in the chemical industry was entered into in 1990. These agreements are basically in the nature of a joint statement of intent, and do not introduce any significant changes in the procedural rules provided by the legislation. However, a growing number of enterprise agreements today deal with technological change; although many of them concentrate on substantive issues arising in the process of change, some (for example, the 1989 agreement at Renault, called "Accord à vivre") have established new rules for labour-management interaction on technology-related issues (for example, union-management joint consultative machinery on training and career development at Renault).

Another French Act of 1982, as amended in 1986, has opened the way to a new channel of workers' influence on issues related to technological change, by granting a right to individual workers to express their opinions, in small group meetings, on the content, organisation and conditions of their work. The Act makes it incumbent upon employers to take the initiative in negotiating an enterprise agreement with representative unions on the modalities of the operation of such group meetings. A recent government report shows a significant rise, between 1985 and 1989, in the number of enterprises where such agreements are in force; today, about 65 per cent of enterprises with trade union delegates are covered by such agreements.[14]

A noteworthy development that has taken place in the 1980s in some of the countries just mentioned is the recognition by legislation or collective agreements of the right of unions or other workers' representatives to be assisted by outside experts in their dealings with employers on technological issues. In France, for example, a legislative reform in 1982 enabled the works councils in enterprises employing 300 workers or more to call on an expert, at the expense of the employer, to assist them in examining management plans for technological innovation. In Sweden, unions have been able to include provisions in co-determination agreements that enable them to call on a consultant, also at the expense of the employer, for assistance in examining management's rationalisation plans.[15]

Mainly collective bargaining

In such countries as Italy, the United Kingdom and the United States, the main sources of procedural rules for labour-management interaction on technological innovation are collective agreements. Union efforts to influence the process of technological change have been focused on obtaining, through collective bargaining, the right to be informed of management's plans for changes in advance so that negotiations can be started early enough to be effective.

In Italy, the right to information, which is today stipulated in collective agreements in most industries, covers a wide area of enterprise policies, including production prospects, investment programmes, restructuring, employment prospects, mobility and training, technological and organisational changes. The Italian labour relations system does not normally make technological innovation a specific subject of negotiations, but rather places it within the overall process of enterprise restructuring,[16] while the labour relations systems of the United Kingdom and the United States tend to deal with technological innovation separately from other aspects of enterprise policies.

Thus, the 1977 national agreement in the Italian printing industry provided for enterprise-level bargaining on work organisation, employment levels, retraining and mobility programmes, and the 1979 national agreement established the right to advance information on investment and restructuring programmes at enterprise level.[17] Our case-study on the Italian banking industry also shows that the 1976 national agreement introduced the right to advance information at the enterprise level, and the 1980 agreement strengthened this right.

In the United Kingdom, the impetus given by the TUC check-list has led to the spread of technology agreements, that is, "formal collective agreements which directly and explicitly attempt to establish union influence and involvement over the process of technological change and its effects on work and the conditions of employment".[18] It is widely estimated that no more than a few hundred such agreements have been signed, mostly at a very decentralised (either enterprise or establishment) level, and the vast majority by white-collar unions, which indicates that only a small proportion of British workers are covered by technology agreements.[19]

Most technology agreements have fallen short of the union objectives set out in the TUC check-list. A survey conducted in the early 1980s revealed that only 8 per cent of agreements stipulated that change would be by mutual agreement (as the TUC had called for), while 50 per cent provided for consultation prior to the introduction (in most cases, implementation) of changes.[20]

In the United States, the inclusion of a requirement of advance notice of the introduction of new technology into collective agreements, which is listed as one of the priority elements of the model contract language of the IAM, still seems a rare practice. Although the International Typographical Union (ITU) local union in Chicago, which we studied, has succeeded in incorporating such provision, the IAM has largely failed to do so. According to two recent surveys carried out by the IAM and quoted in our case-study, none of the 71 IAM agreements in the tool and die industry or the six area-wide IAM master agreements contain provisions for advance notice of technological change.

In Japan, also, collective agreements may be regarded as the basic source of procedural rules for labour-management interaction on technological change, in that it is mostly through them that joint consultative machinery, which deals with technological change among other issues, is set up. The conclusion of agreements specifically dealing with the introduction of new technology, like the Nissan "memorandum of agreement on the introduction of new technology" of 1983, is a rare occurrence in Japan. According to a survey carried out in 1982, 59 per cent of unions were parties in joint consultative bodies, of which 77 per cent had been established by collective agreements. Of the total number of joint consultative bodies, 43 per cent were competent to deal with the introduction of new technology.[21]

2.1.2 Methods of solving problems

The methods of labour-management interaction for solving problems arising in the process of planning and implementing new technology often differ from those for establishing procedural rules for such labour-management interaction. The main methods are joint project groups, joint consultation and collective bargaining.

In Sweden, labour-management interaction on the introduction of new technology predominantly takes place within joint project groups, established within enterprises either under local (that is enterprise-level) co-determination agreements or more informally without any written agreement. Our Swedish case-studies provide a vivid and detailed description of the workings of the joint project groups. In essence, they are study groups within which management and union representatives cooperate in search of the ways of introducing a particular technological change that are most propitious to the achievement not only of a high level of efficiency but also of a good work environment, stimulating work and secure employment.

There is, however, a significant variety in the structure of joint project groups as well as in the ways they operate, as our three case-studies clearly show. The small commercial bank Bohusbanken set up a project group composed of two management representatives and one shop steward. In contrast, the Gothenburg establishment of the large ball-bearing manufacturer Svenska Kullager Fabriken (SKF) adopted a three-tier structure: a policy-making steering committee at the top, an establishment-level project group, and four development groups responsible for looking into problems related to new technology, work organisation, training and work environment respectively, and making proposals. Both workers and management are represented on all these bodies. In SKF, researchers also participate in the steering committee and the project group, and individual workers participate in development groups. It is noteworthy that the participation of individual workers in development groups is in line with the provisions of the Efficiency and Participation Agreement (UVA), the private sector central co-determination agreement, which stresses the desirability of individual employees participating directly in co-determination.[22]

Another notable feature of the Swedish system of labour-management interaction on new technology is the role played by the Development Programme for New Technology, Working Life and Management. The Programme was launched in 1982 by the central organisations of unions and employers, in collaboration with the Swedish Working Environment Fund, and subsidises study projects in various industries aimed at improving job content and work organisation, as well as productivity and competitive capacity, in conjunction with the introduction of new technology. In the SKF experience, the Development Programme subsidised the recruitment of researchers who assisted the steering committee and the project group; it also sent its own representatives to the steering committee.

In Germany and France, joint consultation between management and elected employee representatives (*Betriebsräte* and *comités d'entreprise*) is the main method of problem-solving during the process of technological change, but collective bargaining also plays a significant role. Joint consultation in Germany through works councils (*Betriebsräte*) differs significantly from the Swedish system of labour-management cooperation through co-determination and project groups in several respects. First, German joint consultation does not involve individual workers in decision-making on technological changes as Swedish project groups often do. Many German enterprises have introduced small group activities (for example, quality control circles), but this has mostly been done by management unilaterally. Although works councils, which at first rejected small group activities, have today increasingly come to accept them, German unions still seem to be somewhat suspicious of such activities.[23] Second, German joint consultation on new technology is basically a process of negotiating a social plan aimed at protecting workers from the negative consequences of technological changes through financial compensation and establishing criteria for the selection of workers to be dismissed (section 112, Works Constitution Act of 1972). It is not instrumental in influencing technological innovation, although works councils are today increasingly attempting to assume this new task.[24]

Collective bargaining both in Germany and in France also deals with substantive issues arising from technological change. For example, a number of rationalisation protection agreements and the 1987 Volkswagen technology agreement contain provisions on job protection, training and income protection for workers affected by technological change. A few collective agreements between the Metal Workers' Union and individual enterprises (for example, Volkswagen and Vögele) also attempt to influence the ways in which new technology is to be introduced, by shaping the wage payment systems in such a way that workers' skills are enhanced as a result of technological changes. French inter-occupational and industry agreements on new technology tend to be basically in the nature of a joint statement of intent, often lacking concrete provisions. However, a growing number of enterprise agreements today deal with specific problems arising out of technological changes, such as those in Citroën and Renault on the adaptation of job classifications to the introduction of new technology (1987 and 1989 respectively).

In Japan, too, joint consultation is the main method of labour-management interaction for problem-solving in the process of introducing new technology. Apart from setting up the framework for joint consultation, collective bargaining seems to play only a minor role, although a few agreements (for example, a 1974 agreement in the port industry and the above-mentioned 1983 Nissan agreement) deal with job security and, in the case of the latter, working conditions, safety and health, training and transfer. According to a survey carried out in 1983, the introduction of microelectronic equipment was dealt with by joint consultative machinery in 26.1 per cent of enterprises, while it was dealt with in collective bargaining only in 1.5 per cent of enterprises. The role of both procedures in technological change decreases as the size of the enterprise becomes smaller. Among enterprises employing less than 300 workers, more than 50 per cent had introduced such equipment without any consultation.[25]

As mentioned earlier, 43 per cent of joint consultative bodies in Japan are competent to deal with the introduction of new technology. Among such bodies, 30 per cent provide that the employers must explain to the union plans for technological changes and answer questions; 24 per cent provide that they must explain the plans and listen to the opinions of the union; 39 per cent provide that they must consult with the union, although in the event of failure to reach agreement they may implement the changes unilaterally; and 8 per cent give the union the right to veto the introduction of changes.[26]

The effectiveness of formal joint consultation seems to be relatively limited, if it is to be measured by the percentage of cases of technological innovations in which management change their initial plan after consultation. According to a 1984 survey, it was 6.4 per cent with respect to transfer, 8.6 per cent with respect to manning levels, and less than 10 per cent with respect to most other subjects.[27]

On the other hand, informal consultation seems to play an important role. Union representatives often try to influence decisions through informal discussions with management. In addition, informal consultation often takes place between the first-line supervisors and individual workers, and many decisions on the implementation of new technology are taken at this level. According to the 1983 survey, in 21.7 per cent of enterprises, the introduction of new technology was dealt with in workplace discussion meetings involving first-line supervisors. Under these circumstances, formal consultation sometimes tends to become a mere formality through which decisions taken informally at the shop-floor level are approved by union and management representatives.[28]

Informal consultation is not peculiar to Japan. It is extremely important in many countries, particularly among small and medium-sized enterprises. In Germany, for example, informal consultation between management and works council reportedly contributes significantly to the defusion of the potential for conflict in reconciling the parties' interests in the process of technological and organisational changes.[29] This observation seems to be corroborated by the findings of our German machine-tool case-study.

In Italy, the United Kingdom and the United States, collective bargaining is the main method of dealing with substantive problems arising in the process of introducing new technology. It is set in motion either periodically or ad hoc as a result of an advance notice of changes given by management. There are, however, differences between the three countries, and they are growing in certain respects. In Italy the relatively broad coverage of the right to advance information, mentioned earlier, has led to the development of collective bargaining on a wider range of subjects related to new technology than in the United States or the United Kingdom. Thus, following the conclusion of an agreement in 1977 at Olivetti on investments in new technology and on employment levels and work organisation after technological change, agreements on similar subjects were concluded in a number of large firms.[30] Our newspaper case-study also shows the important role played by collective bargaining in the process of introducing photocomposition, and, more recently (in 1987), in the experimental introduction of a direct entry editorial system.

In the United Kingdom and the United States, on the other hand, the available information indicates that relatively little consultation and negotiations are taking place between management and union over the introduction of new technology. In a survey carried out in 1984 in 2,000 workplaces in the United Kingdom, manual shop stewards reported that only in 27 per cent of cases had they been consulted by management before it was decided to introduce microelectronics technology; in 12 per cent of cases union officers had been consulted. In the same survey, management reported that, in only 6 per cent of the establishments that had recently introduced microelectronics technology had management negotiated with union representatives over the introduction of that technology. Even when analysis was confined to places recognising unions, changes were negotiated in only 8 per cent of cases.[31]

In the United States, unions have traditionally focused their efforts, in the area of technology, on defending the jobs of union members by defining precisely, through collective agreements, the rights attached to each job, and ensuring, through grievance procedures, that management respect these rights. This has been true of the International Typographical Union (ITU) and the International Association of Machinists and Aerospace Workers (IAM), which were the subjects of our case-studies.

The effectiveness of collective bargaining as an instrument for job protection has, however, varied widely between the two unions. The ITU local union in Chicago has largely succeeded in securing exclusive jurisdiction over the new composing room work that accompanied technological change and over new composing room technology, and in including in collective agreements provisions giving the union control over the speed of technological innovation. The IAM, on the other hand, in spite of its effort to enhance worker control over technological change through its model contract language, has made few collective bargaining inroads with respect to technological change in the machine-tool industry. According to the two recent IAM surveys quoted earlier, none of the 71 IAM agreements in the tool and die industry or the six area-wide IAM master agreements contained provisions preventing lay-off as a result of technological change.

In the three countries that rely on collective bargaining for dealing with substantive problems related to technological change (Italy, the United Kingdom and the United States), the practice of joint consultation (meaning labour-management cooperation for solving problems in contrast with negotiations for setting standards) is limited, at least within enterprises in the three industries we studied. The Italian banking case-study reports that a standing union-management consultative committee was established in 1985 to deal with all aspects of the bank restructuring process, but its activities were irregular and suspended after two years. The British banking case-study refers to the existence of a Joint Standing Committee on New Technology between management and one of the unions in the bank, but it was suspended in 1987 during a pay dispute, and therefore did not provide opportunity for discussion at a major stage of technological innovation.

Our case-studies contain several references to industry-level consultative machinery for dealing with technological changes. However, even such machinery tends to function as negotiating bodies or grievance machinery, as is the case with the Joint Standing Committee set up by the ITU local union and the Chicago Newspaper Publishers' Association.

Recently, there has been a significant trend towards the development of joint consultation in a number of enterprises even in these countries, such as car manufacturers in the United States and state-owned enterprises belonging to the IRI group in Italy (for example, a joint technical committee responsible for training and retraining made necesssary by technological change, set up in Italtel – the largest Italian producer of telecommunication equipment – through the enterprise agreements of 1981 and 1982). However, the trend still does not seem conspicuous in the three industries studied.

2.2 *Impact of labour-management interaction on the planning of technological change*

This subsection seeks to assess the extent to which workers and their representatives have been able to influence the planning of technological change, namely decision-making on whether or not to introduce technological change, the selection of technology, the extent of technological change, and the speed of introduction of new technology. Workers may influence the planning of technological change, thus defined, not only directly by participating in decision-making on it, but also indirectly by influencing the consequences that the planned technological innovation is allowed to have. For example, union success in obtaining a guarantee of job security in the process of introducing new technology might constrain managerial decisions on the extent and speed of change as well as the type of technology to be introduced. Likewise, their success in obtaining a guarantee that workers' skills will be upgraded after the new technology is introduced might influence both the selection of new technology and the methods of its introduction. This subsection focuses on the first type of influence, that is, workers' direct influence on the planning of technological change. Workers' indirect influence will be discussed in the next subsection, which deals with labour-management interaction on some of the consequences of implementing technological change.

Workers' direct influence on the planning of technological change still seems very limited in most countries, but there are nevertheless some areas in which unions in some countries have been able to exert a certain influence. It is in Sweden that workers seem to have been able to exert the most far-reaching, and direct, influence on the planning of technological change. The Swedish model of technological innovation involves the establishment of joint project groups, typically set up under co-determination agreements. Our printing case-study shows that most of the recommendations of the joint project group concerning the selection of new printing presses were implemented by management. The case-study on the ball-bearing producer, Svenska Kullager Fabriken (SKF), also shows the effective influence of joint project groups on the extent of the application of a flexible manufacturing system (FMS) in the newly designed production systems, while the banking case-study shows strong union influence on the extent of technological change. It is noteworthy that, in both the bank and the printing firm, the initiative in introducing new technology came from workers. It is also noteworthy that, in one of the largest commercial banks in Sweden, to which our case-study makes reference, workers' influence extended beyond their participation in joint project groups, and was exerted also through a joint administrative committee responsible for decision-making on technical, administrative and organisational changes.

The effectiveness of workers' influence varies with enterprise, however, even in Sweden. The available evidence seems to indicate that the existence of a system of co-determination and joint project groups on technological change does not by itself guarantee the success of labour-management cooperation on this issue. It seems also necessary that management should recognise the contributions that workers can make to the successful introduction of changes and be willing to involve them from the planning stage. Workers and their unions must also be equipped to propose adequate alternatives to management plans. The successful experience of the Malmö establishment of Scanian Dairies (Skanemejerier), referred to earlier, fulfilled both conditions. In cases in which labour-management cooperation has been practised only in compliance with the Co-determination Act, workers reportedly tend to complain that the information provided by management is so scanty and so belated, and management so unwilling to listen to the unions' views, that the subsequent negotiations do not have any real effect.[32]

Another difficulty confronting Swedish workers and unions involved in joint project groups lies in the decentralised nature of work in such groups. As such groups are normally small and operate at a decentralised level, union representatives in them have difficulty in understanding the consequences of proposed changes for the whole work process and for the entire workforce in the enterprise; they are also aware of the risk that such a decentralised process may be used by management to create divisions between different groups of workers and different categories of employee.[33]

In the other countries we have studied, workers' direct influence on the planning of technological change seems generally to be significantly less, although the printing industry presents a picture that is somewhat different from this general pattern. Unions in the printing industry, in particular in the newspaper industry, have exerted considerable influence on the speed of introduction of direct input advertising or editorial system in many countries, in particular during the earlier stages of the changes. Our British printing case-study, for example, refers to a technology agreement between the National Graphical Association (NGA) and the management ensuring the gradual introduction of the direct input advertising system. However, unions in this newspaper later lost influence when management proceeded to the introduction of the direct input editorial system. The NGA also temporarily succeeded in preventing the introduction of direct input in *The Times* and other national and provincial newspapers towards the end of the 1970.[34] Our American printing case-study reports that, for each technological innovation that occurred in the Chicago newspapers, the publishers determined unilaterally and without union consultation which type of technology was to be deployed, but the ITU local union retained the power of veto over the introduction of new technology.

In Italy, the newspaper *Corriere della Sera* provides one of the few examples, in the private sector, of union involvement in decision-making on the introduction of new technology from a very early stage. The enterprise agreement of 1985 deals in detail with technological investments, specifies the number and type of new machines as well as the time of their introduction, and provides for periodic joint control of the process of change.[35]

In Japan, on the other hand, a recent case-study on a major newspaper presents a picture of a weaker union influence on the planning of changes. In this enterprise, management has consistently regarded the introduction of new technology and new equipment as a matter within the province of managerial prerogative, on which it has been willing to inform the union but not to consult with it.[36]

Workers in other industries seem to have been unable, in most countries, to influence the speed of technological innovation as effectively as printing workers. More generally, their influence on the planning of technological change seems to have been weak. In the United Kingdom, in spite of the TUC's concern with influencing the design of technology, new technology agreements have mostly been confined to job security, payment for change and guarantees of union membership. More generally, it is reported that unions have had little influence on investments in new technology or decisions concerning the type and extent of technology to be introduced, although they may have had some influence on the speed of introduction of new technology.[37] In the bank that was the subject of our British case-study, management regarded all aspects of technological change as an integral part of management responsibility, and consequently there were no negotiations over the decision to introduce new technology, the choice of technology, or the speed of technological change; nor were there negotiations on the consequences for employment, job classification, pay systems and other working conditions, or on compensation for any adverse effects arising from new technologies. Another recent study of another bank also shows an unsuccessful attempt by the main union in the industry (the Banking, Insurance and Finance Union – BIFU) since 1979 to seek consultation and negotiation over technological change. Generally, technology is reportedly not negotiable in British banks.[38]

In the United States, direct union influence on the planning of technological change seems virtually non-existent in the machine-tool industry. The die manufacturer studied introduced technological change without informing, let alone consulting, the union. In Italy, union involvement right from the planning phase of restructuring and technological innovation has been experimented with in some public sector enterprises, in particular those within the IRI group, but they are rare in the private sector.

Our German case-studies also give a picture of limited involvement of works councils in the decision to introduce new technology and the choice of technology, although the rationalisation protection agreements in some industries give works councils a virtual veto right or the right to refer disagreements to arbitration, as mentioned earlier. In the banking industry, a survey found that in 83 per cent of banks decisions regarding the introduction of electronic data processing were made by management; only in 7 per cent of cases were works councils allowed to participate in such decisions.[39]

In Japan, a survey carried out in 1983 showed that management only explained their plan for the introduction of new technology to the union in 34.8 per cent of cases, they explained the plan and heard the opinion of the union in 49.2 per cent of cases, and they modified their plan after hearing union views in only 4.8 per cent of cases.[40] In the enterprises we studied, too, union influence on the decision to introduce new technology, the choice of technology, and the extent of technological change seems to have been limited. The machine-tool study, for example, points out that the choice of new technology has never been altered in any significant way because of requests from the union.

2.3 *Impact of labour-management interaction on the consequences of introducing technological change*

Workers' influence on the consequences of introducing new technology seems much more effective than their direct influence on the planning of change. In concrete terms, labour-management interaction on the consequences of introducing new technology covers such issues as how to protect the jobs of workers affected by technological change, how to arrange their grading and retraining, and how to organise work processes. This last issue, work organisation, is discussed separately in a later section, because of its importance as the key linkage between technology and labour relations.

2.3.1 Job security

One of the most preoccupying effects of microelectronics technology is its labour-saving effect. Unions in all IMEC countries (industrialised market economy countries) have been attempting to protect the jobs of their members by various contractual and other arrangements. Such arrangements vary significantly with country and with industry within the same country, reflecting differences in business trends, traditional employment practices, union power and structure and labour relations systems.

To start with *industry characteristics*, our case-studies show that far-reaching employment guarantee clauses have been included in collective agreements in the printing industry. This partly reflects the traditional strength of typographers' unions: in many countries, typographers' unions have enforced *de jure* or *de facto* "closed shops", and exerted an exclusive control over labour processes in the composing room for many decades. The far-reaching employment guarantees also reflect the extent of the threat that new technology represents to the job security of typographers. The introduction of the direct input editorial and advertising system may in theory, lead to the total elimination of compositors' jobs. It is therefore quite natural that typographers' unions in many countries should have focused their efforts on employment protection.

Thus, the ITU of the United States and its local union in Chicago won lifetime employment guarantees against technological displacement for the existing workforce in the mid-1970s. The German printing industry concluded in 1978 a collective agreement on the introduction and use of computerised editorial systems, which contained various clauses meant to protect the jobs of printing workers, for example, a clause providing that, for a period of eight years, editing and correcting work should be carried out only by printers.[41] In the United Kingdom also, as mentioned earlier, the NGA won similar temporary arrangements from *The Times* and other national and provincial newspapers towards the end of the 1970s. In Sweden, a technical agreement entered into in 1974 prohibited dismissals due to new technology. In Italy, the management and the Delegate Council of the newspaper *Corriere della Sera* had by 1973 already concluded a technology agreement whereby the unions agreed to the introduction of new technology in return for guarantees of employment.

In the other two industries, clauses in collective agreements on job security (if any) are generally far weaker. In the banking industry, in particular, there has been little negotiation on employment protection in the context of the introduction of new technology; management generally sticks to its prerogative in making decisions concerning the level of employment. The British bank we studied refused to negotiate on the consequences of new technology for employment, although it unilaterally guaranteed that there would be no compulsory redundancy. The efforts of one union (BIFU) to negotiate a technology agreement failed.

The relative weakness of employment protection provided by collective agreements in the banking industry is partly due to the relatively weak union organisation in banking, which is characteristic of predominantly white-collar employment. However, it also reflects the relatively high job security that bank employees have traditionally enjoyed, as banks in many countries have applied the policy of lifetime employment for many decades. Moreover, banking employment has generally been expanding, or at least stable, until today, in spite of the introduction of new technology, because of the expansion in the volume of transactions.

National differences in the ways in which the consequences of technological innovation for job security are dealt with are also significant, reflecting the differences in prevailing employment practices and labour relations systems. The extent to which previously existing methods of protecting job security have changed in the course of the massive introduction of new technology during the last decade also varies significantly.

In Japan, Sweden and the United States, there does not seem to have been any significant development of special arrangements for protecting job security in the context of the introduction of new technology; ordinary protection provided by legislation, agreements or customs is applied to employment protection against the labour-saving effect of new technology.

The chapter on Sweden refers to the existence of an understanding among unions, employers and the Government that technological or organisational development should not affect the job security of workers; if this does happen, management should provide the workers affected by the change with retraining to enable them to transfer to other workplaces, and if an inter-firm transfer is inevitable, governmental institutions should help in retraining.

Consequently, although a technology agreement has been concluded in the printing industry, special protection against job losses has not been seriously negotiated in either of the firms studied.

Likewise, in Japan, the basic principle underlying the "lifetime employment" system, that workers (at least regular employees) should be retained within the enterprise until their normal retirement age irrespective of business fluctuation, has reduced the need for special measures to protect employment when new technology is introduced. In all the enterprises we studied, and others, management have avoided committing themselves not to resort to dismissals, but in practice have never done so, although the banking and machine-tool case-studies refer to the practice of transferring (or "dispatching") redundant workers to affiliated companies.

In the United States, the far-reaching employment guarantees against technology-induced job losses obtained by the ITU seem to be the exception, according to our case-study. Most IAM collective agreements, and those of many other unions in other industries, contain provisions for laying off workers in reverse order of seniority, which is the traditional method of employment protection used by American unions. Our case-study shows that there has been no change in lay-off procedures as a result of the introduction of new technology.

Nevertheless, it is noteworthy that the spread of what is commonly referred to as "concession bargaining" in the United States in the early 1980s has often involved improvements in job security as a *quid pro quo* for concessions by unions on wages. Thus, for example, American Airlines introduced lifetime job guarantees to existing workers and lowered the pay of new recruits during the 1983 negotiations. The 1982 and 1984 agreements at Ford and General Motors did not provide for such employment guarantees, but they did provide for income support and redeployment for workers displaced by technological change or outsourcing (contracting work *out to another plant or firm*).[42]

In Italy, although there do not seem to have been significant developments of negotiated arrangements specifically aimed at protecting workers against technology-induced job losses, the State has intervened in some cases to facilitate workforce reductions without recourse to compulsory redundancies. This is the case, for example, with the printing industry, where a law was promulgated in 1981 to provide a range of economic assistance to workers made redundant through technological or organisational change, such as income supplements, early retirement and special compensation. Our case-study on *Corriere della Sera* shows the intensive recourse made to the law in reducing the number of composing-room workers by about 100 between 1981 and 1986; 70 of them opted for early retirement, as provided by the law. Our studies generally suggest that the principle of avoiding compulsory redundancy, and resorting to early retirement, recruitment freeze, voluntary redundancy and transfer, is fairly well established.

The development of new negotiated mechanisms for protecting employment when technological and organisational changes are being introduced has been discernible in Germany and the United Kingdom, although its extent seems quite limited. In Germany, several rationalisation protection agreements, for example, at Volkswagen and in the public service, have precluded the possibility of dismissals due to technological change. However, a large majority of such agreements do not preclude dismissals but only relegate them as a means of last resort. The machine-tool manufacturer we studied is not covered by any such agreement, but there is a well-established consensus between management and works council that technological change should not lead to dismissals.

In the United Kingdom, new technology agreements have provided some protection against job losses as a result of technological change. The most frequent provision is that any job loss will occur by voluntary redundancy and natural wastage, thereby eliminating the threat of compulsory dismissal of existing workers.

2.3.2 Job classification and income protection

An important factor determining the degree of acceptability of new technology from the workers' point of view is whether its introduction involves downgrading, and consequent loss of income. However, the significance of job classification as a labour relations issue varies considerably from one country to another.

In Japan, for example, this issue is not a subject for negotiations in the context of the introduction of new technology because wages are linked not to the jobs that workers carry out, but rather to their seniority within the enterprise. Consequently, changes in workers' tasks, which may result from the introduction of new technology, do not normally have any consequence for their grade or pay. This is why our Japanese case-studies do not reveal any special arrangements to protect workers from downgrading.

In those countries where pay is directly linked to the type of job workers perform, on the other hand, possible downgrading as a result of technological change is a serious labour relations issue. Downgrading may result either from the degradation of the job performed by the worker or from the displacement of a worker from his previous job and his transfer to a new one. Some collective agreements cover only the second type of situation, while others cover both.

In the United States, the grading of workers affected by technological change seems still to be dealt with through the application of the traditional seniority rule. This means that management are basically free to downgrade workers due to lack of need for particular jobs, provided they do so in compliance with the seniority rule and the "bumping" system, according to which the most junior (that is, the newest) worker must be downgraded first, and the most senior last. Thus, in the die-making firm we studied, which employed about 50 workers, nine employees in the bargaining unit were reassigned to lower job classifications in the occupational wage hierarchy between 1981 and 1983. Although this was caused by recession, and took place before the introduction of numerical control, technology-induced redundancies would have been dealt with similarly.

The situation is quite different in the printing industry, where the ITU has traditionally recognised only two classes of workers in union composing rooms, journeymen and apprentices: all journeymen, regardless of their specific job assignments in the composing room, receive the same wage. Here, there is no room for downgrading. Even in this industry, however, the recent attempts by employers to obtain the right to make mandatory transfers of composing-room workers out of the composing room into other departments may open the door for their possible downgrading.

In Germany, as mentioned earlier, under most rationalisation protection agreements, management must offer an "equivalent" or "reasonable" alternative job to a worker displaced by new technology. This implies that, if an "equivalent" job cannot be found, the employee can be downgraded and has to accept an otherwise "reasonable" job. Our printing case-study shows that workers who were either unable or unwilling to be retrained to use the new equipment had either to accept transfer to less well-paid jobs (downgrading) or to leave the firm. On the other hand, the machine-tool case-study shows consensus between management and works council that transfers may not involve downgradings. Some collective agreements have established pay systems that eliminate the possibility of downgrading as a result of technological change. The agreement between the Joseph Vögele AG in Mannheim and the Metal Workers' Union of 1983, for example, links the wage to the skill qualification of a worker, thus cutting the link between jobs and wages.[43]

In Italy, as our banking case-study points out, an outright downgrading of a worker seems to be in conflict with article 13 of the "Workers' Statute" of 1970. In practice, there has not been any downgrading, at least in the bank and the newspaper we studied. In the newspaper, there have been negotiations on the regrouping of various composing-room jobs into one photocompositor's job, but this has involved upgradings of some previous jobs, and no downgrading. Both the Italian case-studies show extensive bargaining activities (either formal or informal) on the classifications of new jobs created by the new technology. All the available information points to the importance attached by Italian trade unions to obtaining well-structured, precise job classifications as a means of protecting workers' professional standing.

In neither the Swedish enterprises nor the British bank we studied has the introduction of new technology resulted in any downgrading. The available information suggests that, in general, Swedish parties in labour relations would try their best to avoid the deskilling of workers affected by new technology, while parties in the United Kingdom may tend to place emphasis more on preventing any deterioration in the terms of employment than on job content itself.[44]

In the British newspaper we studied, management agreed to avoid downgrading for existing employees, but applied new wage rates for new employees in new jobs, thus creating two wage rates for the new jobs: one for transferees and one for new employees. Moreover, the NGA members transferred to the editorial room maintained their wage rate, which was higher than the journalists' rates, but their wages have been frozen until the journalists' wages catch up with them.

2.3.3 Training and retraining

Technological change usually involves changes in job content, making many traditional skills obsolete and creating a demand for new types of skill. New technology can yield its optimal benefits only if the workers assigned to new jobs are adequately qualified to carry out the jobs and work with the new equipment. Consequently, training and retraining workers in new skills has become a major priority in managerial policies on technological innovation in all the IMEC countries. Trade unions also regard training and retraining as the most effective means of protecting the employment of workers affected by technological change and other structural changes.

Given such convergence of interests among managers and workers, it is no surprise that there is today wide consensus that training is a subject on which cooperation between labour and management is possible and beneficial.[45] In spite of the existence of such consensus, however, our case-studies suggest that union involvement in training and retraining still remains limited in most countries. In many cases, union influence on training is limited to demanding that management should train existing workers rather than recruiting workers from the external labour market. In a significant number of cases, unions influence the selection of workers to be trained. In a very few cases, unions actively participate in the formulation of training programmes.

It is again in Sweden that workers' influence in this area seems the strongest among the countries we have studied. Co-determination agreements normally provide for union participation in decisions concerning training. All the three Swedish case-studies show that unions have played an active part in both the formulation and the implementation of training programmes, the main machinery for such union involvement being the project or development groups mentioned earlier.

The German system also enables the works councils to exert a significant influence on training, more on the ways in which existing training programmes are to be implemented than on the formulation of such programmes. The Works Constitution Act of 1972 gives the works councils the right to make proposals on the organisation of training and to be consulted on it, but the employers are free to take their own decision. Most rationalisation protection agreements provide for training only in those cases "where the employees can be offered another job ... requiring retraining".

The works council can, however, exert a stronger influence under certain circumstances. First, under the Works Constitution Act, if there are already training programmes in the establishment, the works council has a right to co-determination on the way they are implemented; this means that consultation may lead to arbitration in the event of persisting disagreements (for example, on the selection of trainees), or to a decision by the Labour Court if the disagreement is about the choice of trainer. Second, in some enterprises, technology agreements make it incumbent upon the employers to arrange training in all cases of technological change in agreement with the works council, as is the case, for example, with the Volkswagen technology agreement of 1987.

In neither of the two German enterprises we studied does the works council appear to have played a significant role in the organisation of training, except perhaps in the selection of workers to be trained. Some other studies show incidences of works councils influencing employers' decisions on training through the negotiation of social plans in the context of the introduction of technological or organisational changes.[46]

Our Japanese case-studies suggest that workers' influence on training is generally weaker in Japan than in Sweden or Germany. Another recent case-study on a major newspaper also showed that management designed training programmes: they consulted the union on them, but refused to negotiate. In this case, union influence was mostly confined to demanding that all, not only some, composing-room workers should receive training in the use of new equipment.[47] These findings are roughly in line with those of a 1983 survey, quoted earlier, on labour-management interaction on the introduction of microelectronics technology. In 18.8 per cent of cases, management only explained their plans for training to the union; in 47.6 per cent of cases, they explained the plans and heard the opinions of the union, but unilaterally made decisions; in only 6.7 per cent of cases did they modify their plans after listening to the opinions of the union; and in 10.2 per cent of cases, there was no discussion on training between management and unions.[48]

Our case-studies on Italy, the United Kingdom and the United States show generally limited involvement of unions in the retraining of workers affected by the introduction of new technology, with a few notable exceptions such as the American newspaper industry and (to a lesser degree) the Italian newspaper *Corriere della Sera*.

The two American case-studies show a sharp contrast in the extent of union involvement in training for new technologies. In the die-making firm, the IAM local union has not had any direct influence on training. On the other hand, the ITU and its local union have played a very active part in retraining composing workers for the new technology. Indeed, our case-study reports that the local union gained its first retraining provision for teletypesetting in 1949, so the policy of retraining the existing workforce is half-a-century old in the ITU.

Procedurally, the ITU local union has influenced retraining through the establishment of a principle in successive collective agreements that the first opportunities for retraining should be given to journeymen and senior apprentices. In 1971, in order to reduce favouritism in the allocation of training opportunities and to increase such opportunities, a Joint Selection Committee was established to select workers for a 60-day retraining programme. Although the majority of the ITU local newspaper members were trained through employer-provided on-the-job training, a significant percentage was retrained by the local union training school.

In Italy, vocational training has traditionally been regarded as a matter of direct interest for either employers or public institutions, rather than for trade unions. Trade unions have generally paid little attention to the issue of employee training and retraining at the enterprise level.[49] Union involvement in training has also been hampered by the absence of a tradition of labour-management cooperation in Italian industry. In some enterprises, however, there have recently been experiments with union involvement in this area. Thus, for example, in the IRI group, referred to earlier, a joint technical committee was set up by agreement in 1981 in each enterprise belonging to the group to deal with all aspects of training and retraining, including training needs, programmes, teaching techniques and selection of trainees.[50]

The case-study on *Corriere della Sera* also shows that there were informal, constant contacts between the Delegate Council (*Consiglio di Fabbrica*) and management representatives over training under the 1985 agreement. However, it seems that the unions' interest has centred on the effects of changes on the overall level of employment rather than on specific training programmes.[51]

All the British case-studies show a virtual absence of union involvement in the formulation of training programmes or the conduct of training for workers affected by technological change. In the newspaper studied, there have been some negotiations on the selection of workers to be trained, which have resulted in an offer by management of four editorial positions for NGA members. However, management applied its own recruitment procedures, and as a result only one compositor was selected for transfer to the editorial room.

3. New technology and the reorganisation of work

The introduction of computerised machines into the workplace usually makes reorganisation of work necessary. The ways in which such reorganisation is carried out vary widely. We focus on the following aspects of work organisation:

- the skill composition of the workforce carrying out a unit of production;
- the extent to which the direct operation of machines is separated from the planning of work;
- the degree of fragmentation of work roles, and the degree of rigidity of the division of labour (whether tasks and skills overlap with each other);
- the extent to which each work process is controlled by management.

3.1 Technology and work organisation

The relationship between microelectronics technology and work organisation is not straightforward. Indeed, a particular type of technology does not in itself determine the organisation of work, and even when two enterprises introduce the same new technology, they may arrange their work methods differently. There are a number of factors that affect the causal relationship between new technology and work organisation. These factors include the influence of workers and unions, and more generally the influence of the system of labour relations, as well as the types of work organisation existing at the time of the introduction of technological change. They are analysed in the following subsections.

Some of the factors intervening in the relationship between technology and work organisation are not directly related to labour. This is the case, for example, with batch sizes and the size of factories. In workshops producing products in large batches, work organisation tends to become more rigidly segmented and distribution of skills more polarised than in workshops producing products in small batches.[52] Likewise, in so far as the operation of numerical control and computerised numerical control is concerned, small establishments tend more to amalgamate the operating and programming functions than large ones.[53]

3.1.1 Teamwork and polyvalence

Nevertheless, technology is one of the main factors influencing the way in which work is organised. For one thing, it is widely recognised today that, in manufacturing industries, computerised machines tend to yield their optimal results when they are operated by workers working in teams (see, for example, the French metal industry agreement of 1987 on the introduction of new technology, art. 8, para. 3), because one computerised machine is capable of performing a number of production processes that were separate when conventional machines were used. Accordingly, the introduction of microelectronics technology is often accompanied by the introduction of teamwork.

Teamwork encourages polyvalence among workers, not only in manufacturing but also in other industries, including the printing industry (where the introduction of photocomposition has led almost everywhere to the amalgamation of different jobs in the composing room like typesetting and layout, creating new definitions such as "photocomposition process operative" and the banking industry (in particular, customer counselling services and all types of low volume/high value transactions).

3.1.2 Skill structure

Microelectronics technology also tends to affect the level of workers' skills.

The printing industry

In the printing industry, it tends to deskill (if not eliminate) the work of compositors, if the skill level is to be evaluated by the time needed for training. Our Japanese case-study shows that it used to take three to seven years to train a fully qualified compositor, but today it takes only six months to train a fully qualified operator of computerised typesetting machines. A recent study in the United States has analysed the evolution of the typesetting craft skill index, calculated in reference to the number of years spent in training; it also shows a steady decline since the introduction of photocomposition.[54]

However, if we take criteria other than the time needed for training, the effects of microelectronics technology on the skill levels of compositors appear more complex. According to our United States printing case-study, a multifaceted analysis of 29 composing-room occupations, contained in the *Dictionary of Occupational Titles*, suggests that greater deskilling and stratification of occupations occurred in the transition from hand composition to linotype than in that from linotype to computerised photocomposition, and that the change from linotype to computerised photocomposition has involved more a change in the types of skill (from linotype skills to keyboard skills) than in the level of skills.

Work associated with colour printing has also been deskilled in terms of the time needed for training. The Japanese case-study shows that it used to take ten years to train a worker to be proficient in colour separation, but today a worker is fully qualified to operate a colour scanner after six months' experience.

The banking industry

In the banking industry, the influence of new technology on the level of skills of tellers is more ambivalent. According to a recent ILO report,[55] in general, two divergent tendencies can be observed. In routine transactions, certain skills – of a mechanical nature, but nevertheless requiring a measure of mental effort and concentration – are no longer required, or are required to a lesser degree than previously. The skills replacing them are equally mechanical but they call for less mental effort. The level of skills required for the performance of routine transactions therefore actually falls, although the levels of attention and concentration required will be just as high or even higher. In contrast, in the area of customer services, computerisation offers potential for an increase in both the range and level of the skills required, for example in searching for, extracting and assimilating relevant information in response to a request. The realisation of this potential is, however, contingent on the relevant organisational decisions being taken by management.

The metalworking industry

In the metalworking industry, computerisation provides employers with the possibility of deskilling machinists' work and reducing labour costs. Our United States case-study shows that the percentage of high-wage jobs has significantly decreased in the workforce of the die-making firm since the introduction of a numerically controlled (NC) machine tool, and attributes this to the increase in NC machine-tool operator's jobs.

One of the main factors determining the level of skills after computerisation is the extent to which skilled machinists are entrusted with the task of programming the machines. The more it is entrusted to technical engineers or programmers, the more deskilled the machinist's work will become, and there will be a polarisation of skills between those who programme the operation and those who operate the machines in accordance with the programmes produced by others.

However, the programming of NC and CNC machine tools is itself becoming increasingly simple and easy. A Japanese machine-tool firm employing 1,357 workers reported in 1983 that, on average, it took three months to train a worker with no experience to operate NC machine tools, six months to read programmes, and 12 months actually to make programmes, while it took about four years to master the skills needed to operate conventional machine tools.[56] CNC machine tools are even easier to operate and programme.

Thus, apparently, there is a serious risk of skill degradation for skilled machinists. However, the causal relationship between new technology and machinists' skills is far from being straightforward, for several reasons. First, new technology may be introduced in such a way as to retain workers' craft skills. Management seem today increasingly willing to do so because craft knowledge enables the operators of computerised machines to check the adequacy of the programme and improve it in the light of their knowledge of the production process. Moreover, the presetting of computerised cutting machines, for example, the setting of jigs and fixtures, still requires craft knowledge of the cutting work, and it is perfectly conceivable that the operators should be given responsibility for doing this.

Second, although the time needed for training in programming computerised machines is shorter than that in operating conventional machine tools, this does not necessarily mean that the work with computerised machines is less skilled than that with conventional machine tools. This is because the former normally requires a greater mental ability than the latter, as reflected in the predominance, in many countries, of high-school graduates in occupations related to the programming of NC and CNC machine tools, in contrast with the predominance of workers with lower levels of general education in occupations related to the conventional machine tools. These factors make the comparison of the skill levels of these two types of work very difficult.

3.1.3 Working hours

New technology also tends to affect working hours. As investment in computerised machines is often costly, management are incited to prolong the operating hours to recoup the investment. The fact that much computerised equipment is capable of functioning with the minimum of supervision also facilitates the prolongation of operating hours. Consequently, shift-work is spreading among enterprises that have recently introduced technological change, as our German printing case-study illustrates.

The introduction of shift-work has also been extensive in the banking industry in a number of countries as a result of the introduction of new technology. This development is particularly notable because shift-work was previously virtually unknown in the banking industry in most countries. Our Italian case-study shows that in 1970 a national agreement for the industry already existed that empowered the banks to introduce shift-work as a means of dealing with the consequences of technological change. Our Japanese case-study also refers to the introduction of two and three shifts to carry out computer-related work, which is rapidly increasing.

3.2 Labour relations and work organisation

3.2.1 Methods and extent of workers' involvement

Work organisation has traditionally been regarded as a matter largely falling within the province of managerial prerogative. However, as technological change and other structural adjustments have been massively introduced into the workplace, the issue of work organisation has become an important labour relations issue in most IMEC countries. There is, however, a significant variety in the methods and degree of involvement of unions or workers in decision-making on work reorganisation.

In Sweden, Germany and Japan, there are mechanisms in their labour relations systems that facilitate joint problem-solving in reorganising work when introducing new technology. In the United Kingdom, the United States and Italy, on the other hand, collective bargaining, namely a process of making rules, is the predominant method of labour-management interaction.

In Sweden, as mentioned earlier, workers participate in the reorganisation of work through joint project groups. In Germany, labour-management interaction on work organisation, as on technological change, mainly takes the form of consultation between management and works council, within the framework set by legislation and collective agreements. The Japanese approach focuses on the involvement of individual workers through their work groups in decision-making on work organisation; union influence on work organisation generally seems rather limited.

The degree of workers' influence seems to vary with country and enterprise. Our Swedish case-studies point to relatively effective workers' influence, while the German machine-tool case-study presents a picture of consensual relationship between works council and top manager (reported to be not uncommon in Germany), in which informal contacts play an important role.

In Japan, on the other hand, the relative weakness of unions at the shop-floor level seems to reduce union influence on decision-making on such issues as the speed and scale of production, and the size of the work group. However, each work group is relatively autonomous in arranging its work, normally through informal discussions between a foreman (who is often not only a first-line supervisor but also a representative of the work group and a union representative) and members of the work group (individually or collectively). Under these circumstances, formal negotiations between union officials and management in the area of work organisation tend to play a minor role.

Among the countries where collective bargaining is the predominant method of labour-management interaction on work organisation, that is, Italy, the United Kingdom and the United States, it is in Italy that unions have sought to play the most proactive role in work organisation. While British and American unions have focused their efforts on the defence of rights attached to particular jobs in the context of rigid divisions of labour, Italian unions have been seeking since the early 1970s to enhance *professionalità* (professional standing), both by defending the traditional characteristics of skilled work (ability, apprenticeship, technical expertise, etc.) and by recasting the traditional divisions of labour with a view to increasing autonomy, personal discretion, decision-making power and polyvalence.[57] It is also in Italy that we find significant experiments with labour-management cooperation on work organisation, in the industries we studied, for example, the experience described by the case-study on *Corriere della Sera*, as well as a joint technical commission, established in the newspaper *La Stampa* in 1980, to study, and make proposals for, the new organisation of work in the light of technological innovation.[58]

However, generally, in the 1980s, with the accelerating pace of technological change and structural adjustments, the initiative on work organisation in Italy seems to have shifted from trade unions to management, and the reorganisation of work processes is discussed increasingly from the viewpoint of requirements for increasing productivity.

3.2.2 *Influence of labour relations systems on decision-making*

How then do different systems of workers' involvement in decision-making on work organisation affect the final decisions taken by management? It is of course difficult to answer the question because it is difficult to isolate the effects of labour relations from the effects of other factors. In some cases, however, the pervasiveness of certain tendencies notable in certain specific labour relations systems enables us to establish a link between the two.

In Sweden, the operation of joint project groups on new forms of work organisation clearly fosters the creation of broad work roles and the granting of a relatively large amount of autonomy to work groups. The achievement of these results is also encouraged by public policies through the work of the Swedish Working Environment Fund, from which the parties in joint projects often seek assistance. Our case-study on the ball-bearing manufacturer (SKF) shows that, in accordance with advice given by the Fund, the parties in project groups have agreed on creating a work organisation based on semi-autonomous work groups and giving each group considerable technical and administrative autonomy.

Likewise, in Germany, the pressures from works councils for new skill requirements within the internal labour market to be met by means of training and job enrichment tend to prompt management to retain craft skills and use them flexibly in broadly defined tasks, rather than fragmenting and deskilling jobs.[59] Some enterprise-level collective agreements (for example, the previously mentioned agreements that the Metal Workers' Union concluded with Volkswagen and Joseph Vögele AG) have also incited management to do the same.[60]

In Italy, the ambitious strategies of unions for enhancing *professionalità* are reported to have remained largely abstract objectives due to the unions' failure to follow them up with concrete bargaining demands.[61] Nevertheless, there is evidence of significant union influence on work organisation in isolated cases. Our case-study on the newspaper *Corriere della Sera* and another study on *La Stampa* quoted earlier report a successful union effort to amalgamate various functions in the composing room into one broad job after the introduction of photocomposition.

On the other hand, certain features of a labour relations system may hinder the development of broad job definitions. This is what a recent American study suggests. It argues that the seniority rules governing promotion and job assignments are deterring management from assigning the tasks of programming computerised machine tools (NC, CNC and FMS) to unionised production workers, and instead encouraging management to define programming tasks as exclusively white-collar work, outside the union's jurisdiction.[62] This is because the assignment of new tasks to a job that is already within the bargaining unit binds management with a number of bargaining rules and deprives it of freedom of action. For example, management would be obliged to negotiate over seniority rules concerning eligibility for training, selection or promotion to that job.

3.3 Influence of existing work organisation and skill structure

The types of work organisation and skill structure existing in the enterprise at the time new technology is introduced significantly affect managerial choices concerning new forms of work organisation. This subsection discusses, first, the national differences in work organisation that existed before the current waves of technological change hit industry, and, second, the extent to which work organisation has been affected since then in different countries.

3.3.1 Work organisation and skill structure before technological change

There are wide national differences in respect of the degree of rigidity in job definition and the skill composition of the workforce. Again, the contrast between Sweden, Germany and Japan, on the one hand, and the United States, the United Kingdom, Italy and France, on the other, seems striking.

Sweden, Japan and Germany

In Sweden, Japan and Germany, by the time the massive introduction of microelectronics technology started in the late 1970s, there had already been a departure from the Taylorist system of work organisation based on rigid division of work into minute tasks towards a more flexible and broader definition of the job tasks of individual workers.

In all these three countries, towards the end of the 1960s and in the early 1970s, symptoms of workers' rejection of monotonous and "mind-deadening" industrial work became apparent; workers usually showed their rejection passively through the alarming rise in the rates of labour turnover and absenteeism, but they sometimes showed it in more aggressive ways, for example, through wild-cat strikes, as broke out in Germany in 1969 and 1973. This led to a growing recognition among employers in these countries of the need to redesign and enrich work. Thus, from the early 1970s onward, enterprises in these countries introduced radical changes in the organisation of work. Two of the common characteristics of the new forms of work organisation that emerged as a result were much broader job definitions and significant overlapping among different jobs.

While these three countries share a number of common characteristics as regards the developments in work organisation that took place in the late 1960s and early 1970s, there are significant differences as well. In Sweden, employers launched the so-called "new factories" (*Nya Fabriker*) programme, under which numerous establishments started introducing new production systems, such as parallel grouping of the production flow and the creation of independent work teams to replace conventional assembly lines. One of the first arrangements for assembly without a conventional assembly line was introduced into Saab-Scania's gasoline engine factory in Södertälje in 1972, and the most famous example of independent work teams was started in 1974 in the Volvo Kalmar plant.[63] The new forms of work organisation involved expanded job content and wider responsibility for production workers.

In Japan, since the 1960s, management have been placing increasing emphasis on employee involvement and group work. Consequently, before the current wave of technological innovation had affected industry, work organisation was already characterised by a lack of sharp jurisdictional definition of job duties: job rotation was extensively practised, and task performance was perceived as the goal of a work group rather than of individual workers.[64] This feature was further strengthened in the 1970s when the growing aspirations of young workers for a more humane working life became apparent in the context of a serious shortage of labour, and prompted numerous enterprises to introduce programmes aimed at enhancing job satisfaction; in many cases this led to a broader definition of individual tasks and greater autonomy for work teams.[65]

While in Sweden and Japan workers' rejection of the previous forms of work organisation seems to have been the predominant factor prompting employers to redesign work, there seems to have been another factor in Germany that also contributed to changes in employers' policies on work organisation in the early 1970s: their perception of the changing conditions of the world market. They became gradually convinced of the need to replace the Taylorist system of work organisation, which they judged excessively rigid, by a more flexible system, in order to satisfy the growing demand for high quality and for diversification of product types in the contemporary market.[66] As mentioned earlier, German unions also exerted pressure for the creation of integrated jobs (*Mischarbeit*) as well as teamwork among homogeneous, highly skilled workers. The organisation of work based on such integrated jobs is reported to have a long tradition in the German machine-tool industry.[67]

In these three countries, a broad and flexible definition of jobs and a relatively high degree of autonomy for work groups have led to the creation of a relatively highly skilled workforce, in comparison with the countries that will be discussed shortly. However, such work organisation has been in turn facilitated by the abundance of relatively highly skilled workers, as is the case with work groups in Germany, which are composed of relatively homogeneous, skilled workers.[68] It is also reported that the greater autonomy of work groups in Germany has generally brought about a broader definition of individual jobs and greater overlapping among different tasks in Germany than in some other countries, for example, the United Kingdom and France.[69]

The United Kingdom, the United States, France and Italy

Workers' tasks in these four countries used to be (and often still are) defined more rigidly than in the three countries just referred to. This rigid definition of jobs tends to be accompanied by a greater reliance on unskilled workers and a minute planning and control of work by technical departments and supervisory staff, as seems to have characterised work organisation in France until recently. However, the lack of effective apprenticeship to train skilled workers may also foster such work organisation, as one study pointed out when comparing France with Germany.[70]

In the United Kingdom, as in Germany, there are apprenticeships for training craft workers. However, craft workers are concentrated in maintenance jobs and some jobs in unit technology production; most jobs in production are carried out by semi-skilled workers without apprenticeship. Moreover, while German apprenticeships tend to create a homogeneous workforce because of their continuous nature which facilitates a progression from semi-skilled to skilled trades, British apprenticeships are often an instrument to secure the status of craft union members by creating barriers to mobility from semi-skilled into skilled trades. They are thus more linked with the definition of rigid job territories of skilled work than with the broadening of job definitions.[71]

Once rigidly defined jobs take root, union actions (and sometimes the whole labour relations system) tend to be adapted to them and they may in turn become a factor hindering changes. In the United States, where unions grew after the spread of Taylorist work organisation, involving the division of work into minute tasks with a minimum of scope and responsibility left to workers, they focused their efforts on building walls of protection around these many job and wage classifications, controlling access to them by seniority and other rules, and negotiating rates for each job. The barriers thus created among jobs, in particular those between skilled workers and less skilled workers, seem to have restricted the impact of the "humanisation of work" programmes, launched in the 1970s to redesign work and promote job enrichment and job enlargement. Indeed, the goal of job enrichment could not be fully achieved so long as semi-skilled and unskilled workers were systematically excluded from the job tasks requiring the greatest discretion and responsibility.[72]

Work organisation in a typical French enterprise is characterised by a predominance of unskilled workers whose work is minutely planned and controlled by technical departments and supervisory staff. It is also characterised by a rigid definition of tasks, as illustrated by the fact that the system of remuneration in France is closely linked to the work posts (*les postes de travail*) to which workers are assigned, the work posts determining the classification of workers in the wage scales.[73]

In such a system of work organisation, rotation of workers and polyvalence are difficult to introduce because each change of task may in principle lead to a claim for pay revision.

3.3.2 Work organisation and skill structure after technological change

The characteristics of work organisation and skill composition existing before the introduction of new technology affect the ways in which work processes are reorganised after the introduction of new technology.

Various studies seem to show that, in places where the tasks of workers have already been defined broadly and flexibly with much overlapping among them, the reorganisation of work after the introduction of new technology has been comparatively smooth and workers' resistance relatively minor. Moreover, in plants where the skill levels of workers are high, technological change tends to strengthen the tendency towards the integration of planning and production tasks. In other words, management are relatively inclined to entrust shop-floor workers with planning or programming tasks. In the metalworking industry, for example, management tend to assign programming tasks to the operators of NC, CNC or FMS machine tools in plants where the skill levels of workers are already high at the time these computerised machine tools are introduced.

In places where tasks are defined rigidly and narrowly, on the other hand, the process of reorganisation tends to involve far more drastic changes, and workers' resistance tends to be stronger. Moreover, in places where the average skill level of workers is low, management tend to prefer to separate the planning functions from the operating functions clearly and entrust the former to technical staff, thus reducing the work of the operators to the simplest tasks and fostering the trend towards polarisation of skills.[74]

Accordingly, there are marked national differences in work organisation after the introduction of new technology, reflecting the differences existing in work organisation and the skill composition of the workforce prior to the introduction of new technology. Thus, for example, in manufacturing and process industries, there is reportedly a consistently greater use of shop-floor and operator programming for work with computerised machines in Germany, where line and technical expertise are closely linked with each other, than in the United Kingdom and France, where the latter pervasively tends to be split off from the former.[75] While in the United Kingdom and France, the planning and programming function confers white-collar status, blue-collar workers are more extensively used for programming in Germany.

In Japan, the introduction of NC, CNC and FMS machine tools and industrial robots has brought about a further broadening of the job spans of production workers in a large majority of establishments. Japanese metalworking establishments also show a relatively high propensity to assign the task of programming computerised machines to shop-floor workers. Case-studies of eight metalworking plants conducted in 1983 in Japan show that, in six out of eight establishments, a significant part of the programming of computerised machines was entrusted to the machine operators.[76]

In a few cases, all were entrusted with the full range of programming tasks; in one of them, workers were rotating the functions of programming, presetting and operating machine tools approximately every three months. In only one case was programming completely separated from the shop-floor and assigned to technical staff. In a few cases, programming was done in cooperation between technical staff and machine operators.

In Sweden, too, the available information (including our case-studies) suggests that new forms of work organisation emerging in the course of technological innovation involve broader definitions of tasks and greater overlapping of individual tasks in increasingly autonomous work groups. Although it is difficult to isolate the effects of former work organisation from other effects (in particular, workers' influence through joint project groups), it is clear that the types of work organisation and skill structure existing before technological change have facilitated, and made beneficial, the reorganisation of work processes along the lines just mentioned.

In this respect, the situation prevailing in the United States presents a contrast with that in Germany, Japan and Sweden. It is reported that only 14 per cent of machine operators in establishments where computerised machine tools have been introduced have regular programming work.[77] It is also significant that, in the die-making firm we studied, management initially assigned the programming task for NC and CNC machine tools to the engineering department, and workers had to teach themselves programming on the job and convince management of the desirability of allocating some of the programming to them.

4. The influence of technological change on labour relations sytems

The introduction of new technology has affected labour relations systems in various ways. The causal relationship between technological change and labour relations systems is usually indirect, and subject to the influence of a number of intervening factors, including the characteristic features of work organisation and of product and labour markets. However, it is nevertheless possible to identify certain pressures that new technology exerts on all traditional systems of labour relations.

The influence of new technology can be observed firstly in the changing characteristics of the parties in labour relations, in particular workers, trade unions and management, and secondly in the ways in which the parties negotiate and consult with each other. Technological change also has significant effects on the pay system within the enterprise. However, this section does not deal with these last effects, because they will be analysed in detail under another research project of the ILO, now under way.

4.1 Effects on the structure of the workforce

The effects of new technology on workforce structure can be observed in respect of the age and gender composition, the relative numbers of production workers and office workers, inter-firm and intra-firm labour mobility, and types of employment contract. Most of these changes are discernible only in some of the industries in some countries.

The age composition of the total workforce of the enterprises studied has not changed significantly. However, a number of our case-studies have shown that more young workers are assigned to machines incorporating microelectronics technology than middle-aged or old workers, in particular in the printing and metalworking industries. Our Japanese case-studies on the printing and machine-tool industries highlight the plight of older workers in the face of the introduction of new technology. This is mainly due to the fact that older people have greater difficulties than young people in learning the new types of skill needed to operate computerised machines. However, as another case-study on a major Japanese newspaper has pointed out, an additional difficulty for older workers lies in the blow that retraining for new technology represents to their pride, as such retraining is usually given them by young workers with less traditional skills than they have.[78]

The gender composition is changing in banks and some other industries to the detriment of women. In banks, women have traditionally made up the bulk of lower clerical staff. As we have already seen, this is precisely the area of bank employment that has been affected most dramatically by the labour-saving effect of microelectronics technology. Our Japanese case-study shows that the number of women recruited by Japanese city banks has decreased by between 20 and 50 per cent since 1983.

In other industries, the introduction of shift-work as a result of technological change, which has been referred to earlier, is reducing job opportunities for women. A Japanese case-study on an electronic appliance firm shows that, in an assembly unit that was a predominantly female workplace in 1971 (with 60 male and 225 female workers), male workers became largely predominant in 1982 (with 240 males and 115 female workers) after computerisation of assembly and the introduction of two shifts.[79]

The introduction of microelectronics technology has affected the employment contracts of an increasing number of workers. For example, there has recently been a sharp increase in the proportion of part-time workers in the total workforce in a growing number of industries in many countries. In many cases, this is because computerised machines, by significantly reducing the volume of human work needed, make it possible for management to replace full-time employees by part-time employees and so reduce labour costs. The use of part-timers is also facilitated by the relatively short period needed for training workers in the operation of computerised machines, because the work basically consists of keying in instructions, which does not require long training.

The substitution of part-timers for full-time employees has been particularly notable in banks in many countries, where regular female employees have been gradually replaced by part-time workers, most of whom are previous employees returning to employment. A British study reports that the proportion of part-time employees amongst the total employed in the British banking industry has risen sharply since 1981.[80] According to the same study, the increase in the number of part-time workers between 1982 and 1983 was particularly sharp (50.9 per cent) in branches that introduced ATMs during this period, while the number of full-time employees in these branches decreased by 3.5 per cent during the same period. It also shows a growing division between skilled career staff and unskilled non-career staff. These findings are corroborated by our British banking case-study, which shows an increase in the number of part-timers.

Our Japanese banking case-study also shows a significant increase in the number of part-time workers (mostly female) working in data processing, while regular (predominantly male) employees are inreasingly concentrating on work involving a higher degree of discretion and judgement, such as financing, fund operation and consulting services. Most of the female part-time employees in the Japanese case-study, as in the British, are former employees returning to employment after a period of interruption due to marriage or childbirth.

Part-time work is being introduced in the metalworking industry as well. As the operation of NC and CNC machine tools has become increasingly easy, it is possible to assign part-time workers to such tasks, especially in places where programming and operation of NC machines are clearly separated.[81]

For the same reason, management are able today to replace regular workers by temporary workers. The Japanese banking case-study shows that, in the computer division of the bank studied, half of the 200 computer operators are workers sent by temporary employment agencies. This is because the introduction of new technology has temporarily increased data processing work in banks, and many banks have opted to cope with this temporary increase in workload through the use of temporary workers. The Japanese printing case-study also shows the recourse to female workers sent by temporary employment agencies to do copy-typing.

The percentage of white-collar workers in the total workforce seems also to be rising, as some of our case-studies on the metalworking industry show. The Swedish case-study reports a blurring of the demarcation between white-collar and blue-collar employment as a result of technological change. The Japanese and German case-studies also report a decline in the number of production workers and an increase in the number of white-collar employees.

We have also sought to find out whether the introduction of new technology has increased or decreased workers' mobility between enterprises or between workplaces within the same enterprise. The findings do not indicate any straightforward trend. On the one hand, computerisation is creating skills that are largely transferable from one enterprise to another. This is the case, for example, with computer programmers. Many other skills seem also to be losing the specificities attached to each enterprise as work processes are automated and incorporated into computer systems. Consequently, the need for extended experience within a specific enterprise is diminishing in some areas of employment. Thus, our British banking case-study points to an increase in labour mobility between banks in an industry where an internal labour market has traditionally been predominant. Our Japanese printing case-study also indicates growing inter-firm mobility among print workers.

On the other hand, it is clear that a number of enterprises are seeking to strengthen their internal labour market by tying the definition of jobs and opportunities for personal advancement closely to the firm's internal labour market. The development of polyvalence can be instrumental in this because the process of forming polyvalent workers is usually highly specific to each enterprise. This is perhaps partly why our case-studies do not show any general trend towards increased inter-firm labour mobility. Our machine-tool study in the United States actually shows a decline in inter-firm mobility after the introduction of new technology.

4.2 Effects on the unions

The effects on technological change on union membership vary widely with industry and type of union. Among the unions organising workers in the three industries we have studied, the craft unions in the composing room in the printing – in particular newspaper – industry have suffered the most dramatic decline in membership as a result of the introduction of the direct editorial and advertising input system, which has virtually eliminated most former types of composing job.

Thus, for example, our printing case-study from the United States reports that the ITU local union, having accounted for 100 per cent of the Chicago newspaper composing-room workforce until 1985, and having been able to prevent technological change from involuntarily displacing any of its newspaper members, saw its membership decline from 1,200 in 1975 to about 500 in 1979, and to roughly 200 by 1988. Our British case-study also reports a membership loss by the closed shop NGA.

Unions in the printing industry organised at the industry or enterprise level have fared relatively well. Our printing case-studies from Germany, Italy and Japan do not show any significant decline in union membership. Likewise, banking unions in all the countries studied present a picture of stable union membership.

In the metalworking industry, our case-studies suggest that most unions are having difficulties in maintaining the previous level of union membership. The American machine-tool study, for example, shows that, between 1980 and 1985, IAM membership in the United States declined annually, on average, by 8.2 per cent, while manufacturing production worker employment declined by only 2.3 per cent annually during the same period. However, the extent of membership decline varies widely with enterprise and with country, and is often caused more by business decline than by the introduction of new technology.

Technological change is also affecting demarcations between unions. In the printing industry, as some of the compositors are moving into the editorial room as a result of the introduction of editorial direct input, the demarcation between compositors' unions and journalists' unions has sometimes been a conflictual issue. This is the case, for example, in the United Kingdom where this issue led to the signing in 1985 of a national-level agreement between the NGA and the National Union of Journalists (NUJ), providing that any workers transferring from the composing room to the editorial room would become joint members of the two unions, but the NUJ would have bargaining rights for the NGA transferees. Our Swedish printing study also refers to inter-union jurisdictional disputes occurring as a result of the introduction of new technology. Demarcation between unions may also become a problem in the metalworking industry, as shown by our Swedish machine-tool case-study, which reports a blurring of demarcation lines between the tasks of foremen (now production leaders) and those of some clerical workers. In the banking industry, on the other hand, our case-studies suggest that technological change has not created any jurisdictional problem in the absence of craft unions.

The declining membership of craft unions is prompting some of them to merge with other unions. In the process, some unions are losing their craft nature. A prominent example in the industries we have studied is the merger in 1987 of the ITU and the Communications Workers of America (CWA) in the United States.

4.3 Effects on management

Several of our case-studies (for example, Japanese printing and Italian banking) point out that the effects of technological change on management have been greater than those on the unions. These effects are notable mainly in respect of the levels of managerial decision-making and the relative importance of different managerial functions.

There have recently been significant changes in the levels of managerial decision-making in many of the enterprises we have studied, although not all these are caused by technological change. Some reflect the characteristics of the decision-making process concerning technological innovation. In general, decision-making on the introduction of new technology seems to be highly centralised, while decision-making on the subsequent implementation of technological change tends to be decentralised. This is what our Italian banking case-study shows.

The British survey of 2,000 establishments, quoted earlier, also shows that decisions to introduce microelectronics technology affecting manual workers are significantly more likely to be taken by management at a level higher than that of the establishment than decisions to introduce conventional technological change.[82] However, the same survey shows that decisions to introduce microelectronics technology affecting white-collar workers are substantially more likely to be taken at establishment level than at a higher level.

Some other changes in the level of decision-making result from the new possibilities opened by new technology for centralisation or decentralisation of decision-making. Our Japanese banking case-study shows that, after the introduction of new technology, management authority has become increasingly centralised at the head office as new technology has enabled management to reduce the size of branches and increase their number. This has resulted in the weakening of the authority of middle-level management, as control over branches by central management has been strengthened. In the British bank studied, on the other hand, new technology has led to operational decentralisation, involving the overall reorganisation of the bank into five groupings, and increasing responsibility for individual sections within each grouping.

These divergent trends seem to show that new technology itself is neutral in its effect on the level of decision-making, but it does provide management with new possibilities for changing the level of decision-making towards either greater centralisation or greater decentralisation. This is borne out by the findings of our Italian case-study on *Corriere della Sera*, which show a decentralisation of decision-making to intermediate staff and individual divisions between 1975 and 1981, when technology changed from hot-metal composition to photocomposition, while the following phase of technological change in the 1980s under the new ownership was marked by the centralisation of decision-making and a reduction in the autonomy of middle-level management. The findings of our case-studies and other research are more convergent in respect of the effects of technological change on the importance of the personnel function. They indicate that the personnel department has usually played only a minor role in the process of introducing new technology, and that technological change tends to decrease the importance of the personnel function. These are the findings of our case-studies on the British banking and the German machine-tool industries. The Italian banking case-study reports that the personnel and human resource management functions played only a secondary role in the early phase of technological innovation, but their role is growing at the implemention stage.

The British survey of 2,000 establishments also shows that technological change is still largely regarded as a technical matter within which there is no established role for personnel management. It shows, for example, that while personnel management have been involved in the introduction of microelectronics technology in 50 per cent of cases, in only 15 per cent of cases have they been involved in the decision to introduce the changes, and in only 19 per cent have they been involved immediately after the decision to change has been taken.[83]

The reverse side of the coin is that the organisational and planning functions of management have acquired greater importance in the process of technological change in a number of enterprises. This trend is clearly shown by our case-studies on Italian banking and German printing, the latter showing the establishment of a central department of work planning in the context of the introduction of new technology. The Japanese printing case-study reports an increase in the number of technical managers.

Technological change may also affect management structure in other areas apart from the level of decision-making and the relative role of the personnel and other functions. Our Swedish metalworking case-study shows that the number of hierarchical levels of management has been drastically reduced as group work has been introduced in the process of introducing new technology, under the influence of the work carried out by development groups and researchers. The changes have involved, for example, the abolition of the categories of manufacturing engineers and foremen.

4.4 Patterns of negotiation and consultation

The introduction of new technology also affects the ways in which negotiations and consultations are carried out between management and unions or other workers' representatives. However, our case-studies show that the extent to which the previous system of labour relations undergoes transformation in the process of introducing new technology varies significantly from one country to another. They also show that new technology alone does not usually shape the new patterns of labour relations, but rather reinforces the pressures for change in conjunction with other factors.

Our German case-studies show a high degree of stability in the patterns of labour relations within enterprises during and after the period of technological change. Changes have taken place at higher levels, like the spread of rationalisation protection agreements since the late 1960s. However, they have not modified the basic features of labour relations within the enterprise. Instead, they have mainly had the effect of reinforcing the functions of the existing institutions for labour-management cooperation and for the representation of workers' interests, while creating new challenges to workers' representatives.

Our German case-studies show that the introduction of new technology has modified neither the functions of the existing labour relations institutions nor the attitudes of the parties to each other and towards technological change. They refer to the "culture of rationalisation" underlying the cooperative relationship between works councils and management, both parties sharing a common knowledge and evaluation of the technology used in the industry, as well as a common appraisal of the basic economic environment surrounding the industry. Such cultural traits of German managers and workers account for the stability of the pragmatic and rational style of consultation through the period of rapid technological change, either strictly in accordance with the provisions of the Works Constitution Act ("norm-oriented consensual action", as it is called in the case-study), as in the printing firm, or through arrangements made by key persons in the enterprise ("person-oriented consensual action"), as in the machine-tool firm.

Our Swedish case-studies also indicate a high level of stability in the patterns of labour relations within the enterprises studied. With the exception of a few large national newspapers where technological innovation has led to some jurisdictional conflicts, the studies present a general picture of continuing cooperation between management and unions, both parties maintaining autonomy in the formulation of their goals. Although the Swedish system of labour relations seems to have evolved more significantly in the process of technological change than the German system, this evolution has been basically towards the strengthening of the tradition of labour-management cooperation.

The new approach towards technological development, as illustrated by the increasing experiments with joint group work, constitutes a departure from the earlier approach, which relied heavily on the work of experts, and encourages greater participation of the parties as well as individual workers at shop-floor level in the development of ideas and the solution of problems as well as in the search for alternatives to management proposals. In encouraging the participation of individual workers, the Swedish development seems to diverge somewhat from developments in Germany, where workers' representation within enterprises seems to remain the virtually exclusive domain of the works councils, which tend to be suspicious of the development of other more individual channels of representation (for example, through work groups).

Our machine-tool case-study from the United States also indicates that labour relations arrangements have not changed to any significant extent in the course of the introduction of new technology. We have already seen that the clauses on advance notice and joint consultation on new technology in the IAM model contract language have not made any significant inroads into collective agreements. The basic means of employment protection has been the traditional seniority rule and the bumping system. The study shows that, instead of developing new patterns of labour relations, management in the industry are attempting to cut labour costs by wresting wage and benefit concessions from the workers and relocating factories out of the north-east into regions of the United States where wages are lower.

Both the American case-studies indicate that there has been a shift of control over work processes from union to management as a result of technological change. This trend is probably discernible in other countries as well, but the extent of the shift is perhaps greater in the United States than in most other countries studied.

In other countries, a trend commonly pointed out by a majority of case-studies is the decentralisation of labour relations. In places where labour relations contacts have been taking place at both industry and enterprise levels, they have tended to become increasingly enterprise-specific, as pointed out by the British banking study. The Japanese case-studies also generally point to the fragmentation of labour relations in the context of an already decentralised system.

This trend is partly due to the growing inter-firm competition for new technology and for markets. In most enterprises we studied, technological change has been regarded as a domestic issue, and inter-firm cooperation for technological development seems to have been insignificant. In some cases, the competition has been exacerbated by other factors, as has been the case with the deregulation of banking in Japan. Our British case-study refers to the collapse of the Federation of London Clearing Bank Employers; although new technology was not a direct cause of the collapse, it is reported to have reinforced centrifugal pressures in the Federation.

Another notable trend is the growing importance of joint consultation and other problem-solving processes (in contrast with standard-setting processes) in labour relations. Several prominent examples of such development, such as the Italian IRI and the American car industry, have already been mentioned. The findings of a number of the case-studies suggest that similar changes, albeit of a narrower scope, are taking place in a considerable number of industries. Our Italian case-study of *Corriere della Sera* shows that consultation and contacts have grown between specialised union delegates and management in technical committees, and the study on banking also points out growing cooperation. The British printing study, too, suggests the possibility of growing labour-management contact as joint quarterly review meetings have been institutionalised to review the operation of the newly introduced technology.

5. Concluding remarks

This chapter has sought to identify the mutual influence that technological change and labour relations exert upon each other. It has shown that labour relations (namely negotiations, consultation and other forms of labour-management interaction) influence the process of technological change both directly (that is, through the participation of workers and unions in the formulation of plans for changes and in decisions concerning their implementation) and indirectly (that is, by determining the consequences the changes are allowed to produce).

The direct influence of labour relations on technological change is still rather limited; in particular, workers' participation in the planning of technological change is still insignificant in most countries, although the efforts of unions to participate in this area may lead to a growing influence in the coming decade in an increasing number of countries.

On the other hand, the indirect influence of labour relations on technological change has rapidly acquired importance in an increasing number of countries in the last decade. By regulating such issues as job classification, work organisation and pay systems, which inevitably arise as technological change is introduced, labour-management interaction is increasingly influencing the selection of new technology, and the speed and manner of its introduction.

This sort of indirect influence also points to the existence of factors other than labour relations that influence the process of introducing new technology. The most notable of these is work organisation, interpreted broadly and including the skill levels of workers and the contents of individual jobs. This chapter has shown that the pattern of work organisation existing prior to the introduction of technological changes significantly affects the process of change, because the resilience of the existing work organisation largely determines the extent and manner of the optimum introduction of new technology.

Most notable amongst the other factors that influence the process of technological change are market conditions, in both the labour market and the product market. This factor influences the process of technological change largely independently of the influence of labour relations. Although negotiations and consultations on job security may affect the conditions of labour markets, changes in the composition of the labour force in society and the orientation of product markets, for example, remain largely outside the scope of labour-management interaction.

Technological change in turn influences labour relations. It does so mainly in three ways. First, technological change induces changes in work organisation, and then influences labour relations through the development of new forms of work organisation. A notable example is the gradual spread of teamwork and broadly defined jobs. Such trends encourage the development of new channels of worker involvement at shop-floor level, and tend to create challenges to the traditional mechanism of workers' representation. They also tend to threaten the organisational basis of some unions, in particular craft unions, and foster trends towards their restructuring. On the other hand, workers with broadly defined responsibilities, and working in teams, tend to have a greater will and ability to participate in managerial decisions, in particular those related to shop-floor issues.

Second, technological change tends to affect the composition of the workforce and the types of employment contract, by offering new possibilities to management in regard to work organisation and the use of human resources. This in turn will affect the labour relations behaviour of workers and the representational capacity of unions.

Third, the need to deal with new types of problems that arise in the process of introducing new technology encourages the development of new types of labour-management interaction, designed to solve problems through cooperation and participation rather than to establish standards for working conditions and labour-management relations. Moreover, as decisions concerning technological change are, in most cases, specific to each enterprise, technological change tends to foster the development of labour relations at the level of the enterprise. In places where industry or higher-level labour relations have been predominant, this means some decentralisation. Within the enterprise, technological change tends to affect the *locus* of decision-making within the managerial structure. On the one hand, there tends to be a trend towards centralisation of managerial decision-making on the basic issues concerning the introduction of new technology. On the other hand, shop-floor labour management interaction tends to acquire importance as a process for solving problems arising in the implementation of the decisions taken by the central management.

The different national systems of labour relations have dealt with the problems raised by the introduction of new technology differently, and have been affected differently by technological change. The systems that provide for problem-solving through workers' participation or union-management cooperation have been increasingly relying on such mechanisms. On the other hand, those systems in which the creation and application of standards (for example, the conclusion of collective agreements and administration of grievance procedures) have been the predominant features of labour relations are today subject to mounting pressures for change.

Although the findings described in this chapter suggest that technological change encourages trends towards greater similarity in some aspects of labour relations (for example, the development of consultative, problem-solving mechanisms) and work organisation (for example, the development of teamwork and broader definition of jobs), the extent of such trends still remains unclear.

Notes

[1] Peter Dempsey of Ingersoll Engineers, at a conference on FMS in 1983 in London, quoted from J. Child: "Managerial strategies, new technology and labour process", in D. Knights, H. Willmott and D. Collinson (eds.): *Job redesign* (Aldershot, Hampshire, Gower, 1985), p. 120.

[2] David Noble: *Forces of production: A social history of industrial automation* (New York, Knopf, 1984), pp. 44-92.

[3] Rainer Schultz-Wild: "FMS to Group sagyo", in Masami Nomura and Norbert Altmann (eds.): *Nishi-Doitsu no Gijitsu-Kakushin to Shakai-Hendo* (Tokyo, Dai-Ichi Shorin, 1987), p. 120.

[4] See e.g. the case-study on a car manufacturer and a parts manufacturer, by K. Koshiro and H. Nagano, in M. Sumiya (ed.): *Gijutsu-Kakushin to Roshi-Kankei* (Tokyo, Japan Labour Institute, 1984).

[5] W.W. Daniel: *Workplace industrial relations and technical change* (London, PSI, 1987), pp. 184 and 198.

[6] Norbert Altmann: "Shin Gijutsu to Rodo Kumiai", in *Nishi-Doitsu no Gijutsu-Kakushin to Shakai-Hendoo* (Tokyo, Dai-Ichi Sorin, 1987), p. 67.

[7] ibid., pp. 65 and 88; Michèle Tallard: "La Négociation des nouvelles technologies: éléments pour une comparaison de la France et de la RFA", in *Droit Social*, No. 21 (Feb. 1987), pp. 126-7.

[8] See B. Lorentzen: "Union strategies in introduction of new technology in Denmark", in G. Graversen and R.D. Lansbury (eds.): *New technology and industrial relations in Scandinavia* (Aldershot, Gower, 1988), pp. 70-9; B. Gustavsen: "Technology and collective agreements: Some recent Scandinavian developments", in *Industrial Relations Journal*, Vol. 16, No. 3 (Autumn 1985), pp. 34-42 (in particular on Norway).

[9] Jesper Steen and Peter Ullmark: "Participation and technological change in the Malmö dairy", in Graversen and Lansbury, op. cit., pp. 120-9.

[10] The principles of union action, formulated at a symposium of the professional workers' section of the CFDT in December 1980, quoted from F. Sellier: *L'Impact des nouvelles technologies sur l'action syndicale et le système français de relations industrielles* (Aix-en-Provence, LEST-CNRS, 1984), p. 28.

[11] Sten Edlund and Birgitta Nyström: *Developments in Swedish labour law* (Stockholm, Swedish Institute, 1988), pp. 56-8.

[12] W.K. Blenk: *New technologies and collective agreements in the Federal Republic of Germany: The case of "Rationalisation protection agreements"*, paper presented at the Second European Regional Congress of Industrial Relations, Herzlia, Israel, 1987.

[13] *European Industrial Relations Review*, No. 162 (July 1987), p. 22.

[14] Ministère du Travail, de l'Emploi et de la Formation Professionnelle: *Bilan annuel de la négociation collective: 1989, La tendance*, (Paris, 1990), pp. 87-90.

[15] Peter Ullmark: "The Swedish way of introducing new technology", in Graversen and Lansbury, op. cit., p. 84.

[16] Americo Romano: *The role of the parties concerned in the introduction of new technology: Italy* (Dublin, European Foundation for the Improvement of Living and Working Conditions, 1985), pp. 16-17; European Trade Union Institute, Info. 13, *Technology and collective bargaining* (Brussels), p. 35.

[17] Romano, op. cit., pp. 133-4.

[18] R. Williams and F. Steward: *The role of the parties concerned in the introduction of new technology: United Kingdom* (Dublin, European Foundation for the Improvement of Living and Working Conditions, 1985), p. 17.

[19] ibid.; Paul Willman: *New technology and industrial relations: A review of the literature*, London, Department of Employment, Research Paper No. 56, p. 12.

[20] Willman, op. cit., p. 12.

[21] Ministry of Labour: *Saishin Rodo-Kyoyaku no Jittai* (Tokyo, 1984), pp. 58-67.

[22] Edlund and Nyström, op. cit., p. 55.

[23] Altmann, op. cit., p. 83; W. Müller-Jentsch, K. Rehermann and H.J. Sperling: *Labour market flexibility and work organisation: Germany*, paper presented at an OECD Conference of Experts, Paris, 17-19 Sep. 1990, p. 13.

[24] Altmann, op. cit., pp. 84-7.

[25] Ministry of Labour: *Gijutsu-Kakushin to Rodo ni Kansuru Chosa no Gaiyo* (1983), quoted from Sumiya, op. cit., p. 8.

[26] Ministry of Labour, *Saishin Rodo-Kyoyaku no Jittai*, op. cit., pp. 72-3.

[27] Japan Labour Institute: *Microelectronics-Kiki no Donyu to Rodo-Kumiai no Taio* (1984), quoted from Sumiya, op. cit., p. 8.

[28] See e.g., Y. Kuwahara: "Shinbun-Sangyo", in Sumiya, op. cit., p. 141.

[29] Müller-Jentsch, Rehermann and Sperling, op. cit., p. 13.

[30] Romano, op. cit., p. 16.

[31] Daniel, pp. 123 and 127.

[32] Steen and Ullmark, op. cit., p. 121; Edlund and Nyström, op. cit., p. 49.

[33] This paragraph relies heavily on Ullmark, op. cit., pp. 80-8.

[34] R. Martin: *New technology and industrial relations in Fleet Street* (Oxford, Clarendon Press, 1981), pp. 268 and 299.

[35] T. Treu: "Italy", in A. Gladstone and M. Ozaki (eds.): *Labour-management cooperation in training in the context of the introduction of new technology and other structural adjustments* (Geneva, ILO, 1987), p. 249.

[36] Willman, op. cit., and the literature quoted therein.

[37] Kuwahara, op. cit., p. 125.

[38] J. Child: "The introduction of new technologies: Managerial initiative and union response in British banks", in *Industrial Relations Journal*, Autumn 1985, p. 27.

[39] Müller-Jentsch, Rehermann and Sperling, op. cit., p. 13.

[40] Japan Labour Institute, op. cit., p. 8.

[41] Angela Dirrheimer and Bernard Wilpert: *Introduction of new information technology: Case studies on cooperation between management, works councils and employees*, (Dublin, European Foundation for the Improvement of Living and Working Conditions, 1983), pp. 26, 151 and 178.

[42] This paragraph draws on T.A. Kochan, H.C. Katz and R.B. McKersie: *The transformation of American industrial relations* (New York, Basic Books, 1986), p. 119.

[43] Altmann, op. cit., p. 77.

[44] Paul Willman: *Technological change, collective bargaining and industrial efficiency (Oxford, Clarendon Press, 1986), p. 242.*

[45] See Gladstone and Ozaki, op. cit., p. 392.

[46] For example, Altmann, op. cit., p. 85.

[47] Kuwahara, op. cit., p. 137.

[48] Sumiya, op. cit., p. 8.

[49] Treu, op. cit., pp. 213-14.

[50] ibid., p. 230.

[51] ibid., p. 252.

[52] A. Sorge, G. Hartmann, M. Warner and I. Nicholas: *Microelectronics and Manpower in Manufacturing* (Aldershot, Gower, 1983), pp. 95-6; A. Sorge and W. Streeck: *Industrial relations and technical change: The case for an extended perspective* (Wissenschaftszentrum Berlin für Sozialforschung, 1987), p. 9.

[53] Sorge et al., op. cit.; M.R. Kelley: "Unionisation and job design under programmable automation", in *Industrial Relations*, Vol. 28, No. 2 (Spring 1989), pp. 174-87.

[54] D.T. Scott: *Technology and union survival: A study of the printing industry (New York, Praeger, 1987), p. 129.*

[55] Michael Bell: *The teller and the terminal: The effects of computerisation on the work and employment of bank tellers* (Geneva, ILO, 1988), pp. 27-8.

[56] Koyo Shokugyo Sogo Kenkyujo: *Microelectronics no Koyo ni oyobosu Shitsuteki Eikyo ni kansuru Kenkyu Hokokusho* (Tokyo, 1983), pp. 27 and 76.

[57] European Foundation for the Improvement of Living and Working Conditions: *The role of the parties concerned in the design and setting up of new forms of work organisation: Italy* (Dublin, 1985), p. 1.

[58] Romano, op. cit., p. 144.

[59] Schultz-Wild, op. cit., pp. 130-1; Sorge and Streeck, op. cit.

[60] Altmann, op. cit., p. 77.

[61] European Foundation for the Improvement of Living and Working Conditions, op. cit., p. 1.

[62] This paragraph relies heavily on Kelley, op. cit.

[63] S. Agurén and J. Edgren: *New factories* (Stockholm, Swedish Employers' Confederation, 1980).

[64] R. Cole: *Work mobility and participation: A comparative study of American and Japanese industry* (California, University of California Press, 1979), p. 219.

[65] See S. Takezawa et al.: *Improvements in the quality of working life in three Japanese industries* (Geneva, ILO, 1982).

[66] Altmann, op. cit., p. 64.

[67] Ibid. p. 74.

[68] Marc Maurice, François Sellier and Jean-Jacques Silvestre: "La Production de la hiérarchie dans l'entreprise: Recherche d'un effet sociétal (Comparaison France-Allemagne)", in *Revue française de sociologie*, 1979, pp. 331-65.

[69] ibid.; M. Maurice, A. Sorge and M. Warner: "Societal differences in organising manufacturing units: A comparison of France, West Germany and Great Britain", *Organisation Studies* No. 1 (1980), in particular, p. 70; A. Sorge and M. Warner: *Comparative factory organisation: An Anglo-German comparison of management and manpower in manufacturing* (Aldershot, Gower, 1986), in particular, Ch. 13.

[70] Maurice, Sellier and Silvestre, op. cit., Maurice, Sorge and Warner, op. cit.

[71] Maurice, Sorge and Warner, op. cit., p. 76; Sorge and Warner, op. cit., p. 192.

[72] See Cole, op. cit., p. 200.

[73] Sellier, op. cit., p. 6.

[74] For example, M. Maurice, F. Eyraud, A. d'Iribarne and F. Rychener: *Des entreprises en mutation dans la crise* (Aix-en-Provence, LEST-CNRS, 1986), Ch. 2.

[75] Maurice, Sorge and Warner, op. cit., pp. 71-2; Sorge and Warner, op. cit., p. 101; Maurice, Sellier and Silvestre, op. cit., p. 355.

[76] For example, Koyo Shokugyo Sogo Kenkyujo, op. cit., p. 16.

[77] Kelley, op. cit., pp. 174-87.

[78] Y. Kuwahara, in Gladstone and Ozaki, op. cit., p. 285.

[79] K. Koike, in Sumiya, op. cit., p. 214.

[80] Willman, *Technological change, collective bargaining and industrial efficiency*, op. cit., p. 224.

[81] Koyo Shokugyo Sogo Kenkyujo, op. cit., p. 75.

[82] Daniel, op. cit., pp. 78 and 92.

[83] ibid., pp. 108-10.

2

Technological change and labour relations in Germany[1]

*Gert Schmidt**

1. Technological change and the labour relations system in Germany

Technological and organisational change, and the economic, political and social problems related to it, are essentially the result of strategies of rationalisation realised at enterprise level. For an analysis and evaluation of the influence workers' representative bodies have in this area, especially on the processes of implementing new technologies at enterprise level, one must appreciate the constitutional principle of "dual interest representation": the strict division between enterprise-level regulations on the one hand (Works Constitution Act – *Betriebsverfassungsgesetz*) and regulations regarding the bargaining between employers' associations and unions on the other (wage agreements at industry level – *Tarifebene*).

The principle of duality means that the works council (*Betriebsrat*) represents workers' interests at enterprise level and the unions negotiate wage agreements at industry level. The normative framework of labour relations thus has no place for direct union intervention regarding processes of rationalisation at enterprise level. The works council is the decisive workers' institution here. The union's position may be adopted by the works council but is generally transformed in the light of conditions in the individual enterprise. The union representatives at enterprise level – their number relating to the total number of union members in the enterprise – are not a formal bargaining partner within the enterprise, nor do they have any institutional position within the unions themselves. The enterprise is therefore not only the field of realisation of technological innovation, but also the dominant political arena of conflict and bargaining.

Labour relations at enterprise level are regulated by the Works Constitution Act, which provides the works council with rights to co-determination, consultation and information on social and personnel matters, on questions pertaining to the organisation of work and vocational training, and with respect to so-called "economic matters".

However, this legislation does not only impose limitations on the manoeuvrability of management, but also on the scope of the works council. Thus the latter is not allowed to take industrial action but has to refer conflictual matters to

* University of Bielefeld, Germany.

arbitration or appeal to the labour court. Its potential to make threats is thus severely limited. There is no symmetry of power, however. The requirement to observe the peace obligation formally applied to both sides, but materially almost exclusively affects the works council (and the workforce). The works council has no legal co-determination right regarding enterprise decisions on capital investments.[2] It is therefore basically reduced to bargaining on the consequences of rationalisation. On the other hand, it has an obligation to participate constructively in the realisation of technological change within the enterprise.

As far as negotiations over wages and conditions are concerned, the principle of negotiating at industry level and centralising the negotiation process guarantees that wage agreements between unions and employers' associations regarding level of remuneration, working hours and working conditions achieve the status of general regulations for large regions (generally counties/*Länder*). However, the specific negotiations are carried out at enterprise level between management and works council. The general wage agreements by nature leave a good deal of "free play" for adjustments at the level of the individual enterprise.

In fact, the results of wage negotiations at industry level are related to the average unit in the branch. Works councils in relatively successful enterprises therefore have a realistic chance to work out relevant concessions reaching far beyond these results. The possibility of working out such concessions affects the works council's relationship with the relevant union. On the one hand, the union tries to pursue a strategy of unification, which is, however, bound to average conditions in individual enterprises. On the other hand, the works council tries to work out additional concessions related to the individual enterprise that go beyond the provisions of the general wage agreements, using the instrument of specific enterprise agreements.

The principle of the unity of trade union organisations (*Einheitsprinzip*) and the principle of industry unionism (*Industrieprinzip*) have brought about the union monopoly of the workers' representation at industry level and societal level; this union monopoly is regulated by law.[3] Knowledge of the institutional and organisational framework of labour relations is a necessary but not a sufficient condition for an adequate understanding of the "working" of the dual model of interest representation in Germany. The characteristic of negotiations between unions and employers' associations on the one hand and between works council and management on the other is heavily bound up with a specific socio-cultural "arrangement" that has developed historically during the last decades as an element of national political culture.

Especially in large enterprises and against a background of well-practised structures of interest representation, the formal rules governing consultation and co-determination play an important part in stabilising social relationships in the enterprise and are an important factor in easing the management's policy of technological change. According to Vittorio Rieser, a particular "culture of rationalisation" is the ideological foundation underlying the successful strategy of political and social inclusion of works council and workforce as well as management in the German enterprise.[4]

In many cases technological and organisational innovations in enterprises are mediated to the workforce by the works council. And in many cases it is the works council that takes over the main burden of the job of informing the workers. Often it is its job to persuade the workers that the rationalisation process is in the best interests of the enterprise and that rationalisation means stability in the workplace.

The unions have no choice but to accept the relative autonomy of the works council at enterprise level. Usually, successful works council members are strongly firm-oriented. Because works councils adapt union demands and general principles to specific enterprise-level conditions, unions usually have no direct influence on management decisions. This "division of labour" may result in important additional concessions on the part of management in periods of economic prosperity. On the other hand, it gives management and employers' associations a chance to play off works council against union and vice versa. In periods of economic depression this undoubtedly results in a considerable weakening of the unions' power position.

2. Case-study of a printing firm

2.1 The context

2.1.1 The enterprise

The firm is a privately owned, medium-sized enterprise in the printing industry located in a highly industrialised area in the southern part of Germany. It was founded in the early 1930s, when it was mainly engaged in the production of weekend supplements for a large number of different local newspapers (up to 400).

The firm emerged in its present form after the reorganisation of the German publishing industry after the Second World War. It has a particular market orientation in that it unifies text and picture composition. This is, to this day, relatively rare in the German printing industry. Only three other firms in Germany produce text as well as litho. The firm is therefore well prepared for a technical text-picture integration. The firm's most important advertising campaign offers "full service before printing". The firm has a strong position within the market for high-quality text and picture-posters, especially in the realm of advertising. According to a leading member of management, the all-round service has more importance for the firm's market orientation than the specific product.

However, although the enterprise is a market leader regarding typesetting techniques, the litho sector has not been modernised in the same way. In this area, until recently, traditional handwork techniques were dominant.

The economic situation is good. The total business turnover was DM120 million in 1986. This economic success is also reflected in the growing number of employees. In 1986 ten new employees were hired, bringing the total number of employees to 107. In 1986, the composing department had 46 employees and the repro department 25 employees.

During the 1980s, the decision was made to purchase a new electronic colour scanner, as a technical "in-between" step before the introduction of full text-picture integration. At present such equipment is only produced by four firms worldwide, one of which is in Germany. It was from this firm that the new scanner technology was eventually purchased.

The big publishing firms were the first customers for this new technology, but medium-sized firms such as the one under study are now becoming interested. At the end of the 1970s, and against the background of rapid technological development, the firm's management formed an "alliance" (*Interessengemeinschaft*) with five other medium-sized firms in the printing industry. This "alliance" deals with the producers of the new technology as one customer. In addition, the members of the alliance consult each other regarding their experience with the new technology. All six firms in the alliance have a particular market orientation: "high quality of the product". Member firms meet approximately every two to three months. They look forward to exchanging employees in the future.

2.1.2 Labour relations prior to the introduction of the new technology

As a typical skilled blue-collar worker firm in the printing industry, the workforce of the firm is traditionally unionised. The Printing and Paper Workers' Union (*IG Druck und Papier*), a relatively left-wing oriented union within the German Confederation of Trade Unions (*Deutsche Gewerkschaftsbund – DGB*), has always been characterised by the specifically professional consciousness of the compositors or typesetters as a skilled workers (*Facharbeiter*) élite.

The firm's works council has five members. The head of the works council has worked in the firm as a "classical" compositor for three decades and has experienced major technological changes more than once in his professional career. He directly relates the high level of unionisation of the firm's workforce – around 93 per cent, and far above average for the industry – to the experience of technological change, although there have not been any lay-offs. According to him, "You cannot say that the works council is socialist-oriented. On the other hand we do not work 'hand-in-hand' with the management, although we do cooperate".

A general assembly of employees is held every three months, organised by the head of the works council, who talks to the workers beforehand and asks them if there are any problems. The problems are then presented at the assembly in a general statement.

The committee of union stewards (*gewerkschaftlicher Vertrauensleutekörper*) has no political importance in the firm. The centre of action within the firm is the works council. The direct impact of union organisation, including the local union office, in the firm is therefore very limited, although the union representatives are invited to the quarterly assembly. However, the political orientation of the works council assures that union policy is strongly represented within the firm.

Characteristic of labour relations at enterprise level is the mutual "rational" acceptance by management and workers of a situation of diverse political interests. This holds particularly for the consensus regarding the strong union orientation of the works council in the firm. "This is a strike firm", said the head of the works council. "On the occasion of the last strike, 98 per cent of the employees voted for it. This number was higher than the number of union members in the firm. However, we never had difficulties with the management. If the union says that there will be a strike, then there is a strike. The management of the firm accepts it."

Part of the consensus, however, is also a strict orientation of the works council towards the norms of the Works Constitution Act: "Of course, there are matters we have nothing to do with." This orientation towards labour law is a political focus for both works council and management. It provides the basis particularly for personnel policies and policies regarding the social effects of technological change.

2.1.3 General characteristics of employment relationships

The firm's composing department consists of 12 white-collar workers, 31 blue-collar workers and three trainees – 46 in all. The repro department is smaller, with only 25 people altogether: three white-collar workers, 20 blue-collar workers and two trainees.

The qualifications of the employees are mainly based on professional training (*Facharbeiterausbildung*). Quite a few workers have more than one qualification. Due to the technological developments of the past few years, employees in the areas of composing and repro have been confronted with rising qualification demands. Many workers have participated in professional training and retraining programmes. There are definite careers structures within the different departments.

Management and works council agree on the importance of stabilising the employment situation of the firm and in their emphasis on workforce flexibility. The management are highly concerned with the firm's attractiveness on the labour market for highly qualified technicians and skilled workers. The firm had severe problems with staff fluctuation in the repro department before the introduction of the latest technology. Absenteeism is not a severe problem (significantly, the numbers are not known), and there is virtually no remuneration problem within the firm. Salaries are above the regional average, which is itself considerably above the average for the country. All employees are paid by time; there is no piece-work payment.

According to both management and works council, however, the firm does have a problem of significant wage differentials. This problem is based on the industry-related general wage agreement. While the top teams in the composing department earn up to DM27 and more an hour, the regular wage is DM18.47, a differential of 54.6 per cent.

2.1.4 Reasons for introducing the new technology

Even ten years before the new scanner was introduced, it was obvious that the colour printing technology being used at the firm was out of date. Colour scans were being bought in from other firms, and some colour printing business was being turned away. The question was whether to invest in new technology or to continue to rely on subcontracting out to other firms. As the orders for colour printing were increasing substantially, there was a chance to make a new start, but this could be done only with new technology.

Personnel problems in the repro sector provided another strong motivation for investing in the new scanner technology. Qualified employees were tending to leave the firm: a firm can only retain such employees if it has the newest production technology. Two top workers had already left simply because they regarded the firm as too much of a backwater. Furthermore, it is extremely difficult to recruit new qualified employees without the most modern technology.

One important factor in purchasing the particular scanner that was chosen was that the system is compatible with the so-called EBV systems (electronic picture production). The introduction of EBV systems into the firm at a later stage is already being planned.

One great advantage of the new scanner technology is that the data input is done via screen. As the data input can be easily visualised, it is easy to control. The new scanner is also highly economical as it makes it possible to produce four-colour printing in one production process. Finally, electronic screening results in a better quality of product.

2.2 The decision-making process

2.2.1 Decision-making regarding the introduction of the new technology

The decisions regarding investments in new technology by the firm under study were prepared by the top management of the firm in collaboration with the heads of the departments. They had to assume a good deal of the responsibility as they were the most qualified technical experts in the firm.

About three and a half years before the new technology was introduced, it was clear that the repro department was technologically outdated. A business consulting firm was engaged, and their conclusions were as follows:

(1) The firm had to go into the colour sector.

(2) The scanner technology offered the best opportunity of bringing the department fully up to date.

(3) In the future, problems of qualification and training would be critical.

As a first step, the firm bought a small second-hand scanner, which was used to train three workers. The firm then invested in a bigger machine. The whole process went ahead without friction.

Although the firm's middle-range goal is the introduction of full integration of text and picture production by data processing, the original time perspective has, however, proved to be unrealistic, as experts today say that it will be another three or four years before technically perfect systems are available.

2.2.2 The role of negotiation and consultation

The works council was not formally consulted on the decision to purchase the scanner or on the choice of the particular type of scanner. However, the head of the works council was kept informed, from the very beginning, of developments regarding the new technology, although in an informal way. The works council was only formally brought in with respect to the personal and social questions that arose as the new technology was actually introduced.

The problems with which the works council was confronted were, potentially, fairly explosive. The new technology had created an increased demand for qualified workers for the new workstations. On the other hand, workstations were "lost" in the area of traditional montage. One issue facing the works council was the selection procedure for workers for the new workstations. The works council had sufficient influence to ensure that the new jobs were offered to workers already working in the firm, thus providing them with the chance to qualify, instead of recruiting new workers from outside. New recruitment, it was agreed, should only occur if there were no "own" workers for the job. However, there was a limit to this influence: "The actual selection of the workers for training was not our business", said the head of the works council. "This had to be done according to the standards of qualification. There we had no influence." This point is important. The works council did take a strong line over training the firm's workers before resorting to new recruitment, but it did not intervene in the concrete process of selection at the department level.

Another major issue was the introduction of shift work: because the scanner technology was so expensive, it could only work economically if it worked full time. The negotiations between management and works council during this period resulted in written agreements regarding recruitment politics and shift-work regulations (see subsection 2.3 for details of the latter).

2.3 Consequences for the workforce of introducing the new technology

2.3.1 Job security

The total number of employees in the firm has risen during the last few years. It is not possible to say to what extent this is a result of the introduction of new technology. Because the scanner can produce twice as much as the older machines, and enables the firm to do its own colour work, more orders can be accepted and new employees have been engaged. On the other hand, although there were no lay-offs due to the introduction of the new technology, quite a few workers, who were not able or not willing to engage in retraining, had to choose between being transferred to another, less well-paid job within the firm, or leaving the firm. Some left voluntarily.

It does seem that in this firm, as in the printing industry generally, the new technologies, on the whole, are "absorbed" by the workforce of traditionally highly qualified skilled workers. However, problems could arise if screens are introduced in the composing department, as this could cause a considerable loss of workstations in the montage sector.

2.3.2 Work organisation and working conditions

Many compositors have already been forced to adapt to new technology several times during their professional career. The first step was from hot-metal setting to photosetting and lithography, while the next step was differentiated data processing for picture and text. The introduction of computerised integration of text and picture production would be a further professional reorientation within a very short period of time. The traditional compositor was dependent, to a considerable degree, on manual skills, and could not avoid some heavy manual work. Working with electronic typesetting machines requires abstract thinking and basic qualifications in data processing.

The "new" worker is also forced to work in three shifts. While the works council accepted that the new scanner, being so expensive, has to work full time if it is to work economically, there were some problems with the new work schedules. At first, there was one shift from 10 a.m. to 8 p.m., but the works council objected to this. As the workers were not used to shift work, the management had to carry out difficult and protracted negotiations with the works council. Eventually, a three-shift system was formally agreed for the repro department: 6 a.m.-2 p.m., 7.30 a.m.-3.30 p.m., and 2 p.m.-10 p.m.

Another issue was overtime. According to management, if, for a certain amount of time, more than half of one worker's weekly work hours are done as overtime, then it is time to recruit. Despite the engagement of eight new workers, the amount of overtime has not been reduced – something the works council does regard as a problem, although it admits that the firm needs flexibility in order to enable it to handle orders on a very short time-scale. One aim of introducing the scanner technology was to improve working conditions in the repro department. With the introduction of this technology, the department moved out of the basement and into new – daylight – workrooms. The workstations of the traditional compositor and the worker operating an electronic typesetting machine are also different in many respects. The traditional compositor's is a typical "standing workstation", while the worker with the new technology has a "sitting workstation".

2.3.3 *Payment systems and income protection*

The introduction of new technology did not change the existing wage structure, although it considerably aggravated the already existing problem of differentials. The wage structure stipulated in the collective agreement for the industry does not suit the particular conditions of the firm. The job classifications used in the firm in practice do not exactly correspond to the formal wage categories provided for in the collective agreement. Both management and works council feel that they are excessively rigid, and would like to have a bargaining structure that allowed the firm to regulate wage differentials in accordance with the technical and social conditions of the firm.

Only in individual cases did deskilling and downgrading occur during the process of implementing the new technology. Some group leaders were demoted once more to compositors. It is worth noting that all these problems were able to be resolved on an individual basis. In no single instance was there an official conflict between management and works councils.

2.3.4 *Training and retraining*

The firm's decision to introduce the new technology was strongly linked to particular challenges regarding the qualification structure of the workforce and the need to retain and recruit a high-quality workforce. The firm has made great efforts in the area of training and retraining, and both management and works council encourage workers to make use of the opportunities available. Such encouragement is not always successful. Unfortunately, this was the case with the montage workers: some of them had no formal training, the rest had trained as typesetters, and they did not want to engage in further retraining.

Workers in the firm who did accept retraining were trained by colleagues who were already higher qualified, participated in training courses offered by the firms producing the new technology, undertook training courses offered by state institutions or by institutions of employers' associations and unions, or were sent to seminars at various places. Apart from retraining workers who were going to operate the new machines, it was seen as important that those working in the firm's general administration should become more familiar with the new technology.

Finally, in 1985 the firm started taking on apprentices, a decision taken through consensus between management and works council.

2.4 *Effects of the new technology on labour relations*

2.4.1 *Effects on the unions*

The workforce is traditionally almost 100 per cent unionised. The introduction of the new technology did not affect union membership in the firm. The strictly union-oriented works council has cooperated with the management during the process of introducing the new technologies, adopting a critical but constructive attitude. The policy of the works council was early supported by the workforce.

2.4.2 *Effects on management*

Apart from the setting up of a central department of work planning, there were no changes in management structure due to the processes of technological change. However, the positions of those within the management who are mainly involved in promoting the new technology have strengthened. In particular, the firm's training department and the firm's activities aimed at keeping qualified workers within the firm have gained in importance.

2.4.3 *Patterns of negotiation and consultation*

The existing patterns of negotiation and consultation between management and works council were pragmatically adapted to specific situations during the introduction of the new technologies. In general, with the implementation of technological change, the consultation process between management and works council intensified. For example, the number of meetings increased, and problems relating to workforce policies were discussed in more detail than before. As was already the practice, in particularly critical problem areas, such as recruitment policy and work schedules, written agreements were worked out.

There was no union involvement during the process of technological change under study here. On no occasion did the works council have any particularly noteworthy dealings with the local union office.

2.4.4 Conflict

During the process of introducing the new scanner technology, there was no open conflict within the firm. No problems needed to be taken to the labour court or the arbitration board. The process of technological change, on the whole, clearly demonstrates the firm-oriented, pragmatic and critically constructive character of the works council.

2.5 Evaluation

There is no doubt that labour-management interaction during the process of introducing the new technology was successful, as there were no open conflicts (strikes, sabotage, etc.) nor any recourse to juridical modes of problem-solving. Furthermore, neither absenteeism nor the turnover rate of employees increased substantially during the crucial time of change. The labour relations at enterprise level – mainly the interaction between works council and management, on the one hand, and between works council and workers, on the other – did manage the social and individual consequences of technological change. Thus, the rare cases of individual lay-offs, transfer within the firm, downgrading, and so forth, were dealt with either consensually between management and works council or on an individual basis.

Although the unions have not played any formal part in this process, their de facto political influence has been extraordinarily strong in that the works council represents precisely the union position in dealing with problems. This union orientation of the works council is quite accepted by management. However, there seems to be no "fraternisation" between management and works council. A good deal of the apparent political rationality is due to the mutual consideration of the other side's objective interest in the firm.

"Norm orientation", namely a marked reliance on labour laws and other norms, represents one foundation of the political rationality within the firm, in contrast with "person orientation", which prevails in many small and medium-sized enterprises. There is, however, another, deep-rooted action-pattern that was involved in the success of labour-management interaction within the firm in dealing with technological change. Members of management and works council share not only a common general ideological position, referred to by Vittorio Rieser as a "culture of rationalisation", but also a common knowledge and evaluation regarding the technology and the specific economic problems of the industry. This also plays an important part in creating a consensus between management and workers within the firm, particularly in dealing with recruitment and qualification problems.

Last but not least, the success of labour-management interaction is related to the fact that the workforce of the firm, as a whole, is highly qualified and professional. The shared professional work consciousness or *professionalità* of the main body of the workers is traditionally almost exclusive to the printing industry. Resulting from this at the blue-collar worker level is a peculiar matching of the orientation of individual and collective action, emphasising, on the one hand, high individual self-esteem and, on the other, a strong group consciousness as a worker's élite, which is bound up with a considerable potential for conflict if core interests are challenged (as demonstrated by strikes in the printing industry in the 1980s). This same trait of German blue-collar workers implies a positive attitude towards economic and technical rationality, as well as an absence of fundamental ideological opposition to the given capitalistic society as a whole.

This attitudinal configuration is bound up with a high level of identification with the organisation of production at enterprise level and with the "efficiency" of their "own" firm's output. While these characteristics of highly skilled blue-collar workers are present, to some extent, in many industries, they are, traditionally, particularly strong in the mining, steel and printing industries.

3. Case-study of a machine-tool manufacturer

3.1 The context

3.1.1 The enterprise

Until 1986 the firm was a family-owned limited partnership (*Kommanditgesellschaft*). The managing partner was one of the family members. Since December 1986 the firm has been a joint-stock company (*Aktiengesellschaft*). The firm's headquarters and main factory (roughly 700 workers) are situated in an idyllic small town close to the Bavarian Alps in southern Germany. Close to the main factory (approximately 20 kilometers away) the firm operates a branch establishment employing roughly 200 workers. Another establishment with about 200 employees is situated in North Hesse.

The firm was founded in 1920 by five local mechanics. It started by producing fine mechanical instruments for drawing and construction purposes. Since 1950 the firm has manufactured multi-purpose milling and drilling machines. Today, it exclusively produces numerically controlled (NC) machine tools. In 1978, it manufactured its last manual machine tool. The machines and machine systems it produces are best fitted for the manufacture of pieces of small and medium size. The firm is therefore principally oriented towards users of machinery in small and medium-sized enterprises, the so-called "small-business market" – a market that is particularly competitive. The firm exports about 50 per cent of its output; Western European countries and the United States are the major outlets. Among experts with some knowledge of the industry, the firm is well known as one of the most successful machine-tool producers in Germany (about DM200 million total production value in 1986).

In 1976-77 the firm employed roughly 650 workers, but in 1982-83 their number had risen to (roughly) 1,000. By 1986 it had 1,650 employees (including apprentices), which placed it among the larger firms in the machine-tool industry.

3.1.2 Labour relations prior to the introduction of the new technology

Both management and works council accept the "German model" of institutionalised conflict resolution and the regulations governing it. The works council as an institution is fully accepted by the company director and the entire management.

The position of the works council in the firm does, however, depend on the specific function assigned to it by the company director and on the acceptance of this function by both the works council and management. This function consists in the works council performing subtasks of a staff division concerned with company order and social matters. A major task of the works council is to "get things straight", that is, to sort out issues arising in the firm's day-to-day functioning. It thus complements and in part even replaces the personnel department. The assignment of such regulative functions to the works council, which is not uncommon in German firms, requires a high degree of consensus with respect to the use of labour and the general objectives of the firm. The works council also assumes the function of an "ombudsman" and "troubleshooter".

Apart from participating in the settlement of the firm's day-to-day problems, the head of the works council is a member of the so-called Tripartite Board (*Dreiergremium*), an institution whose power is second only to that of the company director as regards general decision-making in the firm. Composed of the production manager, the personnel manager and the head of the works council, the Tripartite Board takes action in case of transfers, displacements, requests for upgrading, and the like. Although the company director reserves the right to take the final decision on any issue, this Tripartite Board must be seen as an important instrument of company management.

The almost daily negotiations between works council and production manager, as well as the works council's cooperation on the Tripartite Board, constitute a variant of co-determination specific to the firm. The works council thus has a say in practically all questions pertaining to the use of labour that come to be dealt with at enterprise level, as well as in all questions involving essential interests of the workers such as transfers, wage rates and so on.

As to the role played by the workers themselves in the politics of the firm, it can be said that their impact on internal labour relations is minimal. They hardly make use of the only institutional possibility they have of voicing their opinion before a larger in-house public, that is, the general assembly of employees. The works council thus holds the uncontested monopoly of representing the workers' interests in the firm.

As already explained in the general introduction to this chapter, trade unions in Germany remain largely excluded from labour relations at the establishment or enterprise level. That this is particularly marked in the firm under study is due primarily to its local structure and "mentality". Another major factor is that it does not negotiate its labour contracts through an employers' association. This is the firm's tradition, and a practice strongly supported by the present director, who is opposed to any "interference from outside". The personal style that is characteristic of the firm's labour relations system is not compatible with external commitment on the part of its actors.

However, management are reasonable enough not to attempt to cut the union out entirely or provide "stumbling blocks" to union mobilisation in the firm. The works councillors are elected on the basis of a list of candidates comprising only members of the Metal Workers' Union (IG Metall). There is thus a clear over-representation of union members in the works council as compared to their representation in the workforce as a whole. Only 40 per cent of blue-collar workers and 25 per cent of white-collar workers are unionised. The overall degree of unionisation in the firm is roughly 35 per cent.

Furthermore, the firm applies the contractual job classification system for assigning workers to job categories and adopts the increases in standard wage rates that are negotiated annually between the collective bargaining agents at industry level. Moreover, it pays actual wages that are considerably higher than the wages paid by other firms in the immediate area, which negotiate their contracts through an employers' association. Given that the firm actually adopts all collective bargaining agreements, and the provisions laid down in the master agreement, the union's demand for the conclusion of an enterprise agreement has so far not met with any positive response from the management.

The state of affairs just described is not enshrined in any written agreement but is founded on "the word" of the company director. The result is that the union influence is virtually non-existent in the firm. There is no body of union stewards in the firm nor has there ever been, according to the various people we interviewed. Certainly, union stewards exist "on paper" at the local branch office of the Metal Workers' Union, but they play no role in the firm. This situation, again, is not uncommon in medium-sized firms in Germany. If one can speak at all of the union's "presence" in the firm, this is almost exclusively mediated via the works councillors who perform tasks of union organisation (for example, collection of membership dues, distribution of union literature, campaigning for members).

There is a consensus between management and works council as to the system of labour relations within the firm. As a system it is relatively free of potential for conflict: it is virtually impossible to conceive of industrial disputes taking place. Likewise, conflictual matters are practically never referred to the labour court. If conflicts do arise, they are conflicts between individual people. The system of labour relations in the firm is referred to by all actors as "style". The term "style" is used to refer to a set of norms and institutions specific to the firm that are not identical with the normative and institutional framework outside the firm (partially supplementing, partially overruling or replacing the latter), but that are accepted by all actors in the firm. These norms and institutions are not regarded as an alternative model to the statutory rules and norms but as the "application" of the latter to the reality of the individual enterprise.

The emergence of "style" as a consensual departure from the norms laid down in the Works Constitution Act may be interpreted as a "deal". In some areas, management does make greater concessions to the workforce (and the works council) than provided for by the law (for example, the firm's principle of avoiding downgradings, or of management giving the works council early and extensive information on the firm's progress, etc.) In turn, the works council, for instance, renounces its right to limit the latitude of management concerning overtime, to refer conflictual matters to the labour court, or to cooperate more closely with the "external" union.

The company director views human relations within the firm in a patriarchical way: he sees himself as head of a family. Apart from the management style practised by the company director himself, the method of "personal leadership" is a principle of the firm.

3.1.3 General characteristics of employment relationships

Traditionally this is a "blue-collar skilled workers' firm". Although the situation has changed somewhat during the last decade, in numerical terms the character of the firm is still very much in line with this image.

The share of white-collar workers has traditionally been relatively small. In 1973, white-collar workers accounted for about 21 per cent of the workforce. Between 1977 and 1982, however, the share of white-collar workers rose to 28 per cent and has risen further since then. The firm's orientation towards manufacturing is also reflected in the fact that a large proportion of white-collar workers are assigned to jobs in the immediate manufacturing area (out of the estimated 260 white-collar workers presently employed by the firm, 160 work in production, 60 each in the construction and quality control departments, and about 35 in operations scheduling). Eighty of the white-collar employees work in the sales department, while less than 30 are concerned with general administrative tasks (personnel management, accounting, etc.).

On average, the qualification level of the blue-collar workers (which account for around 72 per cent of the entire workforce) is fairly high. Roughly 80 per cent of the blue-collar workers are skilled workers, of which two-thirds have the official certificate of proficiency.

From the 1960s through to the 1980s recruitment was a major problem, particularly of skilled labour. The firm nevertheless succeeded in resolving its recruitment problems: it pays *top wages* (the highest in the region), which helps attract people from the neighbouring areas; it has an extensive *in-house apprenticeship training scheme* (apprentices accounted for 17 per cent of the workforce in 1982-83), which is very attractive to young people; it has implemented, assisted by the local Labour Office, specific *vocational retraining programmes* (in the early 1970s, farmers were retrained to be machinists); and it has introduced a number of attractive *fringe benefits*, such as a canteen, free transportation between workplace and home, and an employee pension scheme.

The average age of the workforce is relatively low. According to calculations made by the personnel manager, 300 out of 970 employees are under 30 years of age. According to 1984 data, by the year 2000, only 131 workers will have reached pension age. Labour turnover is virtually non-existent, and sickness figures are low and obviously uninfluenced by economic upturns or downturns: they have amounted to a stable 5 or 6 per cent over many years. The personnel manager explains this fact in terms of the village "social climate" where "everybody knows everybody".

3.1.4 Reasons for introducing the new technology

In the 1970s, the Western industrial societies were facing the most severe economic setback since the Second World War, due to the so-called oil crisis, the general "toughening" and structural change of the world market, the introduction of new technologies, and the widespread national budget crises. In 1975, German machine-tool production began to experience a deep slump, which reached its lowest level in 1977. In 1979, output had climbed again to the 1974 level of production, but by 1980 it had entered a new phase of stagnation. Since then there has been a substantial recovery in the market, but due to the "dollar situation" and international competition on the product market the prospects are unstable.

Against the background of this critical market situation, the strategy the firm's management decided on was to try to reach a successful compromise between the following two goals: achieving and maintaining a technological pace-setting position on the world market, on the one hand, and improving the existing technology, on the other. To follow this strategy it proved necessary to shift from manually operated machines to the new generation of electronically operated and controlled machines. In addition to this, there was a change from single aggregates to integrated machine-systems, the so-called "machining centres". The first step was to introduce the new NC or CNC machinery on the firm's own production line. In the early 1970s, when the firm's product line only included "conventional" machine tools, the firm started to use automatic machinery (especially automatic drilling machines) and NC machine tools in production. The firm saw itself as required to do so as a consequence of the price struggle on the consumer market, the pressure to produce in increased quantity, and the rising demand for higher precision and product quality. Another contributing factor was the shortage of qualified manpower in the firm. It was five years later, in 1976, that the management decided to embark on the actual production of NC machine tools.

3.2 The decision-making process

3.2.1 Decision-making regarding the introduction of the new technology

The fact that the company director is still the firm's "first engineer" is of great importance: the engineering and innovation centre of the firm is the director's office. The decisions made on the introduction of new technologies have therefore traditionally been top-down decisions. Of course, the company director involves his top engineers and technicians in the decision-making process, but the final "yes" or "no" is his.

The company director describes himself as a democratic autocrat, which implies that he hears the opinion of all and then makes and implements his decision. He regards this particular style of leadership as a strategy of survival for his firm: only through prompt and direct action is it possible to achieve the organisational flexibility necessary to meet the changing market demands. As far as the introduction of NC and CNC technolology in the firm's production line is concerned, the company director himself visited the relevant exhibitions and firms and participated in working out the "design" of the implementation, down to the detailed problems at individual workstation level.

3.2.2 *The role of negotiation and consultation*

As already mentioned, the union has no institutionalised part in the negotiations on the introduction of new technology or on internal mobility. The management – that is, the director – started negotiations with the works council once the principal decision to introduce the new technology had been made, after consultations with members of the firm's engineering and research departments. The interaction with the works council was then relatively intensive, dealing with virtually all social and personal problems that arose, down to the individual workstation level.

There was no gap between management and works council in their attitude towards new technologies and the particular challenge for the firm of introducing numerical control. Management and workers had almost identical perceptions of the consequences for the firm's workforce structure and of the social and individual "adjustment problems" facing existing employees.

The problems relating to the introductions of new technology are dealt with by two different types of negotiations: negotiations on wages and negotiations on internal mobility. It is noteworthy that, under German labour law, the influence of the works council regarding technological change is confined to the direct consequences of new technology for working conditions and wages. The works council is not involved in decision-making concerning the technological and economic aspects of changes.

The discussions and decision-making process concerning questions of placing, paying and training workers in the context of the introduction of new technologies are part of the general negotiations in the firm on remuneration and internal mobility. Negotiations on remuneration are generally initiated by the request of individual workers or by the intervention of foremen who are anxious about maintaining the "income balance" in their department. Negotiations on internal mobility are instigated by the company director or other top management members. Demands concerning internal mobility are often related to the introduction of new technologies and generally involve problems of qualification.

Internal mobility occurs at two levels: on the one hand, it occurs as a result of changes in the structure of the workforce due to changing manpower needs in the various departments. Manpower requirements on the basic production line tend to decline, while there is an increased need for manpower in the mechanical engineering department, assembly, quality control, repairs and customer service. On the other hand, the pattern of labour utilisation within the various departments changes. Thus, due to the changes in machinery on the basic production line, workers who have hitherto operated conventional machine tools have to be retrained to operate NC machines, or they must learn to handle the latest generation of NC machines (for example, machining centres). Electricians have to be trained in electronics, and so on. Due to the constant changes both in the firm's production techniques and in its product line, internal mobility is a never-ending process, towards which the firm takes an "experimental" approach.

It is characteristic of internal mobility negotiations in the firm that they take place on an ad hoc basis, that is, when a displacement or transfer is immediately up for decision. The negotiating body is the Tripartite Board. The initiative comes from the production manager, who might want to fill a new post (or a vacancy), and be considering which employees might be suitable. Alternatively, he might want to "lay off" a worker in one department and be pondering which other jobs in the firm could be offered to that worker. The production manager formulates, if possible, several alternatives, which he proposes to the Tripartite Board. Here, initial opinions are formed regarding the preferable solution, including specific training measures.

The production manager then talks to the worker(s) in question and proposes the transfer to a new job, always putting forward as many alternatives as possible. In many cases, the works council also has a talk with the worker to be transferred, in order to get to know the actual objections made by the worker concerned and, if need be, to appeal to him or her to be "reasonable". In case of persistent objections on the part of the worker, the Tripartite Board meets again to decide whether talks should be continued or new proposals made.

When selecting a worker to be transferred, the Tripartite Board must take into consideration the vocational and personal qualification of that worker. There is also a consensus that a transfer may not involve downgrading, although wage losses due to loss of bonuses cannot always be avoided. Nor may internal "lay-offs" lead to dismissals. The works council pays special attention to these last two stipulations. Finally, in principle, the agreement of the worker affected is sought.

The position of the works council is ambivalent here: as a member of the Tripartite Board it takes responsibility for the selection of workers to be transferred. On the other hand, its "special" relationship with the employees is made use of in that the works council is required to "persuade" the worker in case of resistance to transfer on the part of the latter.

The general principles of company policy are formulated in written agreements, but sensitive individual cases are dealt with on the basis of informal "gentlemen's" agreements. The main achievement of the firm's system of internal mobility negotiations is that it minimises the friction involved in internal restructuring through application of the principle of avoiding lay-offs or downgrading. The firm's common practice of renouncing detailed written agreements is both an expression of political stability and a factor in maintaining it.

3.3 Consequences for the workforce of introducing the new technology

3.3.1 Job security and income protection

No lay-offs and no downgrading are two of the principles guiding the firm's labour policy regarding the introduction of new technology. The measures taken by the firm have proved effective on both counts. Virtually no employees were laid off as a consequence of the introduction of new technology. Internal structural problems have been dealt with through internal mobility and training schemes. Although in some rare cases workers had to accept minor losses in income (through the loss of fringe benefits) as a consequence of internal transfer, there has been virtually no case of substantial downgrading or deskilling of workers.

Management and members of the works council are basically in agreement on the beneficial effects of the introduction of NC and CNC technology: overall productivity in the firm has increased during the last five to ten years, and the workers are convinced that technological change will not adversely affect job security in the firm.

3.3.2 Work organisation and working conditions

Comparing the technical differences between the two generations of machine tools, the major trends as regards workforce input and workers' qualifications are: the increasing absolute and relative importance of the non-mechanical section of the production line, and the increasing importance of electronics and NC steering equipment; the increasing interaction between construction, research and development, and production line, at a high level of qualification; and the need for workers with more than one skill in the production line.

Job satisfaction in general has not been adversely affected as a consequence of the changes occurring in the firm during the last years. Labour turnover, traditionally low in the firm, has not changed significantly.

3.3.3 Training and retraining

The introduction of new technologies into the firm's production process has posed a considerable challenge to the firm's traditional recruitment and training policy. The firm was able to meet the challenges of new technologies and changes in market conditions largely because it could draw on a highly qualified labour force inside the firm. The necessity for internal mobility is accepted by the workforce and, where necessary, production workers receive further training.

The firm takes on a large number of apprentices, but there is no guarantee that they will find employment in the firm upon completion of training.

3.4 Effects of the new technology on labour relations

3.4.1 Effects on the structure of the workforce

The use of NC and CNC machinery in production and the subsequent inclusion of NC and CNC machines in the firm's product line led to a change in the structure of the workforce:

(1) The number of workers in the basic production line, which amounted to 130 in 1972, decreased to 82 in 1982.

(2) There was a relative increase in the number of workers employed in the assembly department, as the quantity produced increased. Similarly, the number of workers in the general mechanical department began to rise from the moment that the firm started to produce housings for machining centres.

(3) The situation in the so-called "E-department" changed fundamentally. In 1972, this was a conventional electric engineering shop with 15 employees. In the mid-1970s, a small NC shop was added. Today it is a large electronics department with 70 employees.

(4) There was also an increase in the number of workers in control (60 employees today) and construction (80 employees).
(5) The number of workers in the customer service and repair departments increased, whereas the number in the administrative department continued relatively to decrease.

3.4.2 Effects on the unions

The introduction of the new technology did not change the political climate within the firm nor the general attitude of workers' representatives and workers towards the union. Formal union membership remained stable (about 35 per cent of the total workforce), while the "firm-centredness" of works council politics was reinforced.

3.4.3 Effects on management

There was no change in management and leadership structure, the company director in particular keeping his function and role of "first engineer". However, the firm did run into some problems regarding the recruitment not only of skilled labour but also of technicians and engineers in middle management positions.

It is not easy to attract technicians and engineers to spend their lives in this idyllic rural area. The highly attractive metropolitan areas of Augsburg, Stuttgart, Ulm and Munich are not very far away, but too far for daily commuting. Young engineers and professionals who grew up in the region tend to look for work in one of the metropolitan areas, while people growing up in the latter are likely to stay there.

The introduction of the new technology did not in any way affect the firm's attitude towards remaining independent from the employers' association.

3.4.4 Patterns of negotiation and consultation

The introduction of the new technology did not alter the patterns of labour relations within the organisation; on the contrary, it seems to have stabilised the political infrastructure of the system, the main features of which are:

- Employees exhibit a relatively strong commitment to the firm.
- There is virtually no consciousness of conflicting class interests.
- "Everybody knows everybody." Workers and company director know each other personally, in some cases from early childhood on. This is not to say that these relationships are free from anxiety on the part of the employees, but actual use of disciplinary measures is a rare event.
- Many issues are settled informally at a personal level before they become an official object of labour relations.

3.4.5 Conflict

Conflict management by consensual decision-making and with full participation of the works council is a guiding principle of the firm's labour relations system. The climate of labour-management relations has been cooperative and positive throughout the process of introducing new technology – notwithstanding some cases of clash of interests. Not a single case of legal action is reported. Any conflict that did arise was settled by consensual action within the firm's socio-political system.

3.5 Evaluation

During the 1970s, the firm had to face major changes due to market developments and the impact of new technologies. These include changes in the qualification structure of the workforce, and changes in work organisation. The evident success of the firm in managing these manifold challenges is clearly related to some specific local factors (politico-cultural tradition, labour market situation, etc.).

Of decisive importance for the functioning of the internal system of labour relations, particularly in relation to difficult internal mobility problems, has unquestionably been the existence of a solid basis of consensus between workers' representatives and management, which has developed over years. Indeed, our case-study impressively illustrates a specific model for conflict management, which has come to be known as "cooperative conflict resolution",[5] and which can be found in large parts of German industry. Our case-study shows, moreover, the broad impact – cutting across class antagonisms – of a conception of modernisation, of a specific "culture of rationalisation" that is characteristic of the political culture of German industry.

This model of labour management cooperation does not work simply as the application of a formal normative rule system to the conditions of the individual enterprise, but is characterised by specifically firm-related patterns of consensual action based on standards of "rationality" common to both sides. In evaluating our "case", we therefore have to take up the matter of the particular "style", with its fusion of autocratic elements of decision-making with the principle of delegation and a high degree of works council participation in the firm's decision-making process, with the works council's influence in some areas clearly extending beyond the scope of co-determination as laid down in statutory regulations.

This case-study illustrates that the "personal leadership style" practised by the company director need not be arbitrary or autocratic. The company director has created institutions – such as the almost daily meetings between the production manager and the head of the works council, and of the Tripartite Board – which, despite the fact that the company director reserves the right to take the final decision, have developed a dynamic of their own. The market-conforming rationality, which serves to legitimise the company director's leadership style, at the same time requires him to observe certain rules. He thus cannot easily ignore the institutions created by himself. Moreover, he cannot simply annul, in times of crisis, the existing internal agreements guiding the firm's social policy, such as, for instance, the principle of avoiding dismissals and downgrading in the process of introducing technological changes, which is an established company tradition.

The labour relations institutions in the firm and the specific agreements are tailored to, or have been tailored by, specific persons. Modes of "person-oriented consensual action" (that is, informal arrangements mainly relying on personal understanding among the actors) dominate in many areas of potential conflict over modes of "norm-oriented consensual action". The specific labour relations structure in the firm, which is accepted by both sides, means that there is no dichotomy between "formal" and "informal" bargaining structures.

Another important factor in the success of the firm has been the traditionally high qualification level of the production workers. The skilled labour potential of the firm serves as a "strategic basis" for ad hoc flexibility, adjustment and retraining. Without doubt, the relatively smooth adjustment to the new technologies and changed market conditions has been partly due to the firm's resources of qualified manpower at the shop-floor level.

The model of labour-management relations outlined above functions successfully to the present day. Considered from an external perspective, it does, however, exhibit some in-built problems. This makes it impossible to give an overall positive judgement, much less to recommend the system for imitation elsewhere. These problems are:

- the strong person-centred nature of the labour relations system;
- the considerable utilisation of the works council for "social management" tasks, which certainly clash with its interest representation function;
- the principle of "cutting out the union";
- the individualisation of problems and conflicts as a maxim of the firm's labour policy.

We asked the company director for his opinion concerning the maintenance of stability in the firm's system of labour relations if its major external conditions, that is, stable expansion and business success, no longer existed. If "painful" decisions could not be avoided, he said, he would hope to find sufficient consensus with the works council and the workforce to enable them to be made. In times of crisis, he regarded problem-oriented, flexible, consensual action with the works council as a cornerstone of the firm's policy.

We also asked the head of the works council about the prospects for the firm's labour relations system, especially in case of economic crisis. According to him, the peculiar intermediate position of the works council means that its position in the firm would not necessarily be severely affected by a shift in power relations between management and workforce. The legitimation problems arising for the works council as a result of its precarious role in disciplinary matters would, however, be mitigated by two factors: first, by the principle of avoiding lay-offs and downgrading, which is part of the firm's culture and which makes arbitrary excessive measures against workers unlikely, and, second, through the deliberate regard on the part of management for the concern of works council members to "save their face" vis-à-vis the workforce.

Notes

[1] The former Federal Republic of Germany, referred to as Germany throughout.

[2] Since the reform of the Works Constitution Act in 1972 the works council has rights of consultation and limited co-determination regarding the implementation of new work processes and techniques, in so far as the work process and the workload and stress of workers are severely affected, although the Co-determination Act of 1976 gives no co-determination rights to the works council regarding enterprise capital investment. There is, however, an indirect co-participation in those cases where workers' representatives and union members are members of the Board of Directors (*Aufsichtsrat*).

[3] Mario Helfert gives a very good overview of the institutional-normative conditions of unions' power position regarding technological and organisational change – see *Gewerkschaften und technische Entwicklung* (Cologne, 1987), Ss. 123-148.

[4] Vittorio Rieser: *Kultur der Rationalisierung und industrielle Beziehungen in Italien und in der Bundesrepublik Deutschland*, Arbeitsberichte und Materialien des Forschungsschwerpunktes "Zukunft der Arbeit" (Fakultät für Soziologie der Universität Bielefeld, Bd. 6, 1984).

[5] See F. Weltz, in collaboration with G. Schmidt: *Innovation, Beschäftigungspolitik und industrielle Beziehungen* (London, 1978); short version published in *Statussicherung im Industriebetrieb*, ed. Dohse et al. (Berlin, 1982).

3

Technological change and labour relations in Italy

1. Case-study of a national newspaper

*Aldo Marchetti**

1.1 The context

1.1.1 The newspaper industry

It is necessary to distinguish two separate phases in the Italian experience of the application of new technology to daily newspapers. The first phase consisted in the move from the hot-metal composition process to the cold, using video-terminals and photocomposition; the journalists continued to write their articles on the traditional typewriters and then passed the text to the typographers to type electronically. In the second phase, which leads to the editorial system, the journalists type their articles directly on to the terminal and the typographers do all the non-journalistic work (publicity, entertainment, obituaries, etc.). The Italian daily newspapers, which number about 75, adopted computerised photocomposition in the 1970s and are at the moment moving towards the editorial system.

The first experiments with the new technology began in the provincial press in 1967 (the *Messagero Veneto*, with a circulation of 25,000, being the first), but medium-sized dailies almost immediately followed suit (for example, *Il Messagero* of Rome, with a circulation of 200,000). For some papers the desire to increase the number of pages, with special editions for small cities, but containing the costs, was a major reason for introducing the new computer systems. In the small firms, union opposition was also less. The experience accumulated by the provincial papers was then passed on to the national dailies. Computerised photocomposition thus became a necessity in all plants. Managers wanted to reduce costs by cutting the labour force, increasing productivity, and reducing the space required for producing the paper.

The prospects opened to the printing sector were extremely advantageous. One newspaper calculated that where five people, including typographers and layers-out, were needed with the traditional methods, with photocomposition, at least in theory, only two are necessary.

* Fondazione Regionale Pietro Seveso, Milan.

The first phase of the technological restructuring was accompanied by intense consultative activity and negotiation by the trade unions concerned and the management organisations, both at enterprise level and at industry level. The negotiations over the new technology had already begun at industry level with the collective contract in 1968, which established that the firm had to inform the union organisation before introducing the new technology and was obliged to reduce to a minimum the adverse effects on jobs and skills. Successive collective agreements up to 1973 obtained more specific guarantees on the protection of jobs and the retraining of typographers.

In 1975 a new contractual phase began: confronted with the prospect of journalists using (electronic) keyboards, the typographers succeeded in defending their positions by insisting that the entry of video-terminals into the industry would happen only under terms set out by the typographers.

The Government, for its part, considering the general critical economic climate in the daily press sector, issued Law No. 416 in 1981, designed to ease the introduction of the new editorial system, which provided for special grants and eased credits for technological renewal. It also provided for the early retirement of daily paper personnel after 35 years of activity, and aimed to prevent collective dismissals in the face of the introduction of the new computer technology. Following the introduction of this law, the number of typographers in Italy fell from 14,441 to 12,500 in the period 1981-87.

In the meantime, the average circulation of the counry's daily newspapers increased slightly, the number of newspapers became stable, and the number of pages increased in those national dailies that had recently introduced weekly sections on current topics.

1.1.2 The enterprise

Here we examine the composition and layout departments of the *Corriere della Sera*, the largest daily paper in Italy, based in Milan. At the beginning of the 1970s the *Corriere della Sera* passed from family ownership to the Rizzoli Publishing Group, which in this way became the country's largest vertically integrated publishing company, including paper-making, books, printing and newspapers. As well as the *Corriere della Sera*, which prints 700,000 copies, the same Milan house also prints *Gazzetta dello Sport*, the biggest-selling sports daily, and *Il Corriere Medico*, a low-circulation specialist paper. We will look at the period from 1975 to 1981, when the change from hot-metal setting to photocomposition was being completed.

1.1.3 Labour relations prior to the introduction of the new technology

Labour relations in the *Corriere della Sera* in the late 1960s and early 1970s were marked by a high level of conflict, rooted in a strong tradition of unionism and solidarity, typical of this sector. Ninety per cent of the workforce were union members (the level for the sector is only slightly lower). This combativity had in the past led to salary levels higher than in other industrial sectors and to a high degree of knowledge and ability to control the production process. This led the typographers' union to understand the importance of the computer technology some years ahead of the rest of the industry.

When the Rizzoli Group took over in the early 1970s, it wanted to avoid a direct confrontation with the unions. The decentralisation of decision-making to intermediate staff and to individual divisions fitted in with this strategy. During these years the negotiating system at the *Corriere della Sera* developed and the unions became more participatory in their attitudes.

The union representation consists of a Delegate Council (Consiglio di Fabbrica) composed of 49 delegates representing 1,900 employees (920 white-collar and 956 blue-collar workers), elected every three years by secret ballot. Elections are held for the list of candidates presented by the three sectoral federations, which are affiliated to the three national federations: the Italian General Confederation of Labour, the Italian Confederation of Workers' Unions and the Italian Workers' Union (CGIL, CISL and UIL). There is thus one negotiating body for all workers in the firm, with representatives from the three federations. These have a differing ideological orientation and compete for membership, but are bound by a federation pact that has lasted now for 15 years. The print unions act at two contractual levels – industry and enterprise. The Delegate Council has great autonomy in negotiation with respect to both the territorial and the national unions.

Trade unions and management have always remained separate: no joint committees or other mixed bodies have ever been formed. The workers' representatives and the managerial technical group (see subsection 1.2) have always examined each problem separately and then presented their respective solutions.

1.1.4 Reasons for introducing the new technology

The *Corriere della Sera*, like other newspapers, had accumulated a large deficit in the 1970s due to the rising cost of raw materials and labour and the stagnation in the readership of the dailies. Technological renewal was thus seen as a necessary step to improve the firm's financial position, improve the product, and make it possible to fight off the ever-increasing competition from other papers wanting to take over its first place in the national circulation. Photocomposition seemed to offer faster preparation times and an improvement in news quality with more up-to-the-minute stories.

1.2 The decision-making process

1.2.1 Decision-making regarding the introduction of the new technology

A team was set up by the management to plan the introduction of the new technology. It consisted of the technical director, the divisional chiefs, and the chiefs of general plant, layout and some offices. The management established the editorial line and the investment plan and gave the general directive for the new technology (maximum time and cost). Members of the team (called technical staff) were given full executive powers and so were responsible for the technical decisions (if the management decided to increase the number of pages in a certain paper, it was then up to the technical staff to find the most suitable solution, from the technical and organisational point of view; this was then discussed with the management and later implemented).

In 1975 the management and the technical team decided to replace, gradually but completely, the hot-lead composition process with computerised photocomposition. The change to the new system took place over the three years from 1977 to 1979.

1.2.2 The role of negotiation and consultation

The decisions to introduce new technology and on the choice of technology were taken unilaterally by the management and were not contested by the unions. All phases of planning and execution were accompanied by consultation and negotiation between the parties, however, during which the management informed the unions of the type, cost and possible consequences of the new technology. There was no open conflict, although there were heated discussions. One divergence between management and unions arose over work organisation. The management were of the opinion that the traditional distinction between the functions of typographers (keyboard operators), layers-out and compositors could be maintained, while the unions favoured the unification of these into one new job. On this issue the union view prevailed, and in the end the management were forced to admit that it was the best solution.

The parties to the negotiations were the technical team for the management and a new technology committee for the unions, composed of ten representatives, some of whom were members of the Delegate Council. Negotiations took place at enterprise level, without interference from the Rizzoli Group management, the management's employers' federation or the typographers' union. The provincial representatives of the workers did, however, intervene.

In the process of technological renewal, the Delegate Council of *Corriere della Sera* availed itself of the provisions of the national printing agreement providing for the unions' right to advance information on investments and management strategy. On the other hand, the enterprise agreements on photocomposition, on the professional standing of the typographers, and on the organisation of work became the property of the entire sector with the contracts of 1979 and 1981.

While there was no real joint planning process regarding the new technology, the most important agreements between management and unions did precede the introduction of photocomposition. The agreement of July 1975 established times and means for the introduction of the new technology, ratified the protection of jobs and the rotation of jobs already mentioned, and repeated the joint intention of workers and management to upgrade the product and make it more competitive, and to acquire new newspapers with the intention of using the production time of the new machines to the maximum.

The agreement of January 1976 laid the basis for the retraining of personnel and fixed the criteria for entry to training courses and other technical matters related to the use of labour in production during the period of the courses. Later, the management were to supply the union annually with complete information on any changes it proposed to make in the organisation.

In July 1986 and January 1987, the management presented the unions with two documents on the integrated editorial system it proposed to introduce. There was to be a system of computers and terminals programmed to automate the preparation of the editorial product, including the acquisition and management of text and pictures, and to generate the entire page, including advertising. This system is designed to make possible the reception, selection, correction and memorisation of text and news from internal sources (feature articles) and external onces (agencies, telex, portable terminals, data banks, etc.). The journalists type their articles directly into the terminals and the compositors make up the pages conceptually.

The firm expects to reduce the typography personnel in the preparation departments by about 60 per cent (from 217 to 80 or 90). The unions are not opposed to this plan, but as in the case of the move from the hot-metal to the cold composition process, they are preparing a plan to safeguard jobs and professionalism to the maximum (the plan is hoped to save 20 more jobs than the management propose).

Towards the end of 1987, management and unions reached an agreement on the experimental introduction of the system for the preparation of ten pages of the group's daily sports paper, *La Gazzetta dello Sport*.

1.3 Consequences for the workforce of introducing the new technology

1.3.1 Job security

It was obvious from the beginning that the introduction of photocomposition would lead to a reduction in the number of jobs. At the beginning of this process the number of workers in the preparation department was 320. Initially, this number did not fall, despite the change from the old to the new system and the increase in productivity, the lesser need for labour having been compensated for by an increase in production, that is, an increase in the number of pages in the two dailies. A new popular paper with a circulation of 150,000 copies was also produced at the Milan establishment but distribution was stopped after only two years (1979-81). In the same period the country's biggest sports paper, *La Gazzetta dello Sport*, was acquired and the base was laid for a small trade newspaper for the medical profession.

From 1981 on there was a financial crisis in the Rizzoli Group, which led to the closure of two papers (*L'occhio* and the *Corriere di Informazione*). It was only at this point that the problem of excess personnel was fully realised. In the same year Law No. 416 came into force, providing for the early retirement of typographers and permitting a gradual outflow of redundant workers without open conflict. Between 1981 and 1986 the number of typographers in the *Corriere della Sera* preparation department fell from 320 to 219.

Seventy of these workers chose early retirement, ten found other work, and 20 were transferred to other departments, either because they had higher educational qualifications or after taking internal competitive exams or training courses. In all these cases employees moved to a different job but at the same career level or slightly higher. The firm was thus able to tackle the problem of overmanning without recourse to the redundancy payments scheme or dismissals.

1.3.2 Work organisation and working conditions

Traditionally, workers belonged to one of ten occupational levels, each with a different wage level, depending on their formal qualifications. Under this system, the typographer and layer-out were both on the fifth level and the compositor on the sixth level. Under the new system, proposed by the union and initially not favoured by management, these three jobs were to be rotated and all were fixed at the sixth level. The rotation of jobs was at first opposed by many compositors because they saw it as an attack on the supremacy of their trade. At present a worker changes role every 15 days, the rotation taking place between the roles of compositor, photo-unit operative and layer-out. Usually the compositors carry out their normal role during their shift and, when their own duties are finished, go on to help the layers-out (normally during the last two hours of the shift).

The new technology led to a formal increase in the rhythm of work (originally the agreed stroke rate was 5,500 per hour, now it is 8,500). Up to now, the real rhythm of work has not increased a lot, however.

The negotiations over work organisation, qualifications and professional status were immediately followed by those about working conditions and health. An agreement was signed in 1981 on this topic by the parties in which recourse was provided to a neutral public agency – the Work Clinic (Clinica del Lavoro) – for a study of the possible effects on health of various environmental factors such as lay-out, air conditioning, video radiation and lighting. Following this agreement, a company health group was set up which could call upon the external agency in a consulting capacity; it was composed of two doctors, one senior doctor and five nurses. The group worked on an intervention programme with the environmental commission of the Delegate Council (five members) and the works doctor. The collaboration of all these bodies led to a number of changes in the area of layout, ventilation and lighting. A health book was given to every worker to record periodic health check-ups.

The department has three six-hour shifts, two day and one night. Break times, which are largely informal, have remained much the same (an average of five minutes' break for every hour worked).

1.3.3 Payment systems and income protection

The payment system was fixed by the national collective agreement, which established a different wage level for each of the ten occupational levels already mentioned. Acceptance of the union proposal that the jobs of typographer, layer-out and compositor should be rotated therefore meant an increase in pay for typographers and layers-out, who moved from the fifth to the sixth level. Basic pay was also increased to make up for the abolition of overtime, which had constituted a not insubstantial source of additional income for many typographers. The average pay of the preparation department thus increased, not just because of the amalgamation of jobs but also because of an increase in the number of "protos" or department chiefs, who increased from 12 to 22.

1.3.4 Training and retraining

The Delegate Council played an important part in determining what professional training would be given to those who were to work on the video-keyboards. The management plan was, initially, to send workers on an external course, but the unions asked for and got a company course lasting seven weeks which was mostly, but not exclusively, practical. The instructors were company technologists who had previously taken a course in using the video-keyboards at the manufacturer's establishment. The course lasted six hours each day; it took place during working hours. All relevant workers, from the oldest to the youngest, took this training course. Once the training was over, each worker began a period of approval in production. The course and period of approval were extended in the case of older workers, who were expected to have more difficulty assimilating the new techniques.

During all this period there was continuing contact between the Delegate Council training commission and the managerial staff to discuss individual cases; the delegates from the departments concerned also suggested changes or adaptations to meet the needs of their workers. After the basic course there were two-day refresher courses each time new procedures or technical modifications or the replacement of a machine or video-terminal required it. The long and complicated process of technological transformation also forced the members of the Delegate Council to equip themselves with the extra technical knowledge they needed in order to take part in various commissions (on technology, environment, training, etc.). They travelled to other countries to see the types of solution arrived at there, and took courses on the new computer technology, availing themselves of the right to take training courses of up to 150 hours a year, during working hours, which had been granted to many categories of worker by the national collective agreements.

1.4 Effects of the new technology on labour relations

1.4.1 Effects on the structure of the workforce

While the number of workers in the preparation department fell, there was an increase in the number of electrical technologists, electrical fitters, hydraulic technologists and all other posts related to maintenance. The electrical technologists, who were on the ninth level, bypassed the unions and entered into separate agreements with the management, but all the others went along with the unions.

Although the number of these technical workers increased, a real polarisation of jobs did not occur, however, because all other workers increased their level.

The basic career structure did not change, although, as already mentioned, the number of "protos" or department chiefs increased from 12 to 22. The next step is plant chief, but as there are only four this is a path open to very few. It is a tradition at the *Corriere della Sera* that they are chosen from among the firm's workers, rather than from among the managerial staff. They are selected by the management on criteria agreed with the unions, which include seniority of service and professional competence. Their duties involve supervising all work in their departments.

1.4.2 Effects on the unions

The level of unionisation remained constant throughout the period of transition. The main effect of the technological transformation was on the methods and negotiation models followed by the unions. Since the 1960s and early 1970s, there has been a change in union attitudes, with less emphasis on claims and conflicts (focusing on salary and the qualification system) and more on consultation and discussion. During the period under consideration there was considerable collaboration between the typographers' union and the journalists' union. During the 1970s there was an agreement to share work in the traditional way (journalists on typewriters and typographers on their keyboards), but the traditional separation of interests re-emerged recently with the possibility of the journalists composing directly on the video-terminals with the introduction of the editorial system. But even with the editorial system an agreement has been reached: the journalists to the keyboard – the typographers to all the rest.

1.4.3 Effects on management

An analysis of the effects of the new technology on the management of the Rizzoli Group presents some difficulty because of the changes of ownership in the period under consideration. Their expansionist policy, and other strategic errors, brought the Rizzoli Group into deep financial crisis. At the beginning of the 1980s, the board of management was forced to resign and the firm was taken into receivership.

1.4.4 Patterns of negotiation and consultation

During the transition period between the old and the new methods of production, the degree of conflict at enterprise level decreased, with the attention of the Delegate Council directed towards negotiations on the various consequences of the introduction of the new technology for the workforce.

It seems likely, however, that the introduction of the editorial system will be a more difficult period for the unions. At the end of the period of receivership the Rizzoli Group was bought by a pool of private firms, among whom was Gemina, the financial company of the Fiat Group. The labour relations policy then changed and was brought into line with the Turin motor industry model: decisions were taken at the top and the autonomy of intermediate management was gradually reduced. Union-management contact became less frequent and of less consequence: answers to union demands were delayed because the managerial group had to go to the Group management. With a more centralised model of labour relations, the unions are likely to have fewer possibilities of influencing the firm's plans.

1.4.5 Conflict

While labour relations at enterprise level were cooperative and peaceful during the 1975-80 period, there was at the same time an increase in support for and participation in national strikes for the renewal of collective work contracts and for the introduction of legislation to reform the printing industry.

1.5 Evaluation

The most interesting point about the *Corriere della Sera* in the period 1975-81 is that a process of radical technological change took place without conflict, with an enrichment of the system of labour relations and relative advantages for both parties.

The unions rejected a prejudiced opposition to the new computer technology and a rigid defence of traditional trades and jobs. They gradually adapted their strategy, changing from primarily defending the economic interests of the workers in a conflict-based way to defending the conditions of work, using participatory and consultative methods.

There was a considerable change in the form of representation; initially a political-type body that acted in a unitary way, the Delegate Council began to perform its duties through technical commissions (which also had a negotiating role), making the contact with the management more consistent and the negotiations more efficient and closer to the actual problems of the workers. The number of typographers decreased, but representation remained firmly in the hands of the sector union, with no breakaway closed groups of professional workers or independent forms of representation.

The firm achieved its objective of introducing new technology, which was the basis of its financial revival and an essential instrument in supporting its product against competition. It also obtained a reduction in labour and production costs. All this happened over a longer period than if change had been imposed from the top without negotiations. However, in the latter eventuality, the adverse consequences of technological changes for the employment and skill levels of printing workers would have been much more serious.

It must be added, to complete the picture, that such positive results could not have been achieved without the third element in industrial relations – the State, which made the law that made possible the introduction of the new technology without open union conflict. But, in its turn, this law would not have been passed without the pressure of a movement that comprised typographers, journalists, other categories of publishing workers and the trade union confederations.

2. Case-study of a bank

Tiziano Treu *

2.1 The context

2.1.1 The banking industry

Banks are among the largest users of microelectronics-based technologies. The first and still most widespread type of innovation is the installation of front-office counter terminals: originally mainly single terminals and now more often autonomous

* Catholic University of Milan.

systems of terminals, both back and front office. Rapid growth is also reported in the use of word processors for office organisation and of various instruments for data treatment (personal computers (PCs), dedicated mini-computers, host terminals, etc.). The use of automated telling machines (ATMs) began later than in other countries but is now common even in small banks. A considerable number of banks are in the process of adopting systems of electronic funds transfer at point of sale (EFT/POS). Home banking and cash management are at the experimental stage in a few banks.

The impact of these new technologies on the level of employment is clear. During the 1970s employment in the banking sector grew extremely fast. It increased by over 50 per cent between 1974 and 1984, but growth has since slowed down dramatically, with a 1.2 per cent increase in 1984. The trend is clearer in the largest banks, the most common users of new technologies, which have experienced a net decline in total employment since 1983. Changes in the structure of employment are also visible, although less clearly measured: for example there is already a noticeable shortage of middle-level employees with technical and organisational skills, and in the area of client service promotion.

2.1.2 *The enterprise*

This case-study concerns a medium-sized bank (3,307 employees in 1988), dealing with all ordinary banking operations. In 1988 the bank had deposits of over 4,800 billion lire, and a total patrimony of over 11,000 billion lire. It is a private company, but part of a larger group. The introduction of new technologies at the bank falls into roughly three phases. In the first phase, begun during the 1960s, the electronic data processing (EDP) unit of the Bank was put in charge of centrally processing data received from the various services of the bank. The data were still collected and treated in a traditional way, manually or mechanically. Consequently the impact of the new technology remained limited to the central unit.

The second phase began in 1974-75 with the introduction of teleprocessing procedures. Originally teleprocessing was used only to transfer data from the peripheral units to the central unit, thereby eliminating "paper movements". Subsequently the bank introduced a growing number of peripheral mini-computers capable of processing data locally and interacting with the central computer unit. The number of local terminals also began to grow rapidly. By the early 1980s this procedure came to be applied to basic bank services and operations from the movement of cheques to the operation of personal accounts, deposits, securities, administration, etc.

In the third phase, which has just begun, the installation of terminals is to become general. Moreover, local units and clients will have access to all information available in the central unit. In this phase data processing is diffused and personalised. It becomes more directly functional to the decision-making process of the various units and an integral part of a decentralisation process affecting the entire bank organisation.

2.1.3 *Labour relations prior to the introduction of the new technology*

The pattern of labour relations in the 1970s was on the whole stable and cooperative, with a low level of conflict. This was fairly similar to the situation prevailing in the banking industry as a whole – though possibly somewhat less conflictual and more "static" than the average.

The level of unionisation is high at 78 per cent, in fact higher than the industry average. The industry-based union structure is also fairly common. Three major local unions exist, which are affiliated to national federations; these in turn belong to the three largest Italian confederations: CGIL, CISL and UIL. The federation affiliated to the largely communist CGIL is in a minority position, contrary to the rule in Italy. This is in large part due to the fact that the bank is concentrated in the north-east of Italy, a traditionally Catholic area where the Christian Democratic Party has a strong political influence. This also helps explain the presence of a strong independent union (Autonomous Federation of Italian Bank Employees/Federazione Autonoma Bancari Italiani – FABI), which has a relative majority of affiliated employees. The managerial style during the 1970s, traditional and rather paternalistic, also tended to favour the position of the independent union. A separate union represents the higher levels of employees called "functionaries". It is a "professional" union of a type rather exceptional in Italy. It is traditionally strong both in the banking industry as a whole and in the particular bank analysed, but its role and power are nevertheless minor. The managers, too, are represented by a separate professional union which is nationally organised and bargains nationally. For all practical purposes, we will consider here only the attitude and position of the four unions representing the bulk of the employees.

Within the bank, the system of representation is that of the single union channel, according to the Italian tradition. The practice of electing the shop delegates in general meetings of all employees (functionaries and managers excluded) was introduced in the banking industry, as in the majority of industries, in the early 1970s. As these delegates are all strictly union members, the enterprise Delegate Council is equivalent to a joint committee of the four union sections. These unions, although divided ideologically, thus bargain together. A businesslike attitude is common among the unions in the banking industry, particularly on local issues, and usually prevails over political and ideological divergences.

Collective bargaining takes place at both national and enterprise level. National industry-wide agreements, renewed every three years, establish the basic standards for wages and working conditions. Enterprise bargaining occurs intermittently, usually every two or three years, on matters indicated in the national agreements although with some degree of autonomy. The most common subjects of enterprise bargaining are productivity or yearly bonuses, social welfare provisions, trade union rights and job classification systems. No formal means exist for unions and workers to participate in enterprise decision-making, apart from collective bargaining. But the right of the trade unions to be informed on basic enterprise policies is recognised following national agreements and is usually respected in practice although with a certain degree of informality.

2.1.4 General characteristics of employment relationships

The employment relationship is regulated by and large according to the standards set in national agreements, with a few details agreed through enterprise bargaining. Manpower policies were fairly traditional during the 1970s. Turnover was low. The job classification system is based on ten occupational levels, from the lowest level of blue-collar workers (who are very few) to the highest level of employees, the functionaries, plus three levels of management (these, however, are not bargained collectively).

The pay system is basically related to the levels of classification. However, the wage differentials between the first and last level were compressed during the 1970s due to the egalitarian wage policy adopted by the trade unions in all industries and to the strong influence of the escalator clause (wage indexation), again common to all industries. The first EDP specialists were included in the general classification *system* bargained at the national level and were mostly classified at the lowest level for white-collar workers (second from the bottom of the scale), but the necessity of upgrading EDP employees was soon recognised (see below).

The actual classification of individual employees is fixed bilaterally by management and local unions. Promotions are regulated fairly automatically by seniority up to the level of middle to high-level employees. Job security is well protected by general legislation on unfair dismissals (Act No. 604/1966), and favoured by employment growth. These characteristics imply stability, but also mean a certain degree of rigidity in internal workforce use. This rigidity was not a problem in the fairly protected context of the industry; the pressure of competition was hardly felt in the 1970s.

2.2 The decision-making process

2.2.1 Decision-making regarding the introduction of the new technology

In the first phase – the establishment of a central data-processing unit – the introduction of new technologies appeared relatively simple and non-conflictual: it applied to only one unit of the bank and the possible impact on work and enterprise organisation was not obvious. The decision to introduce the new technologies was taken by the top management of the bank, which also made the consequent choices; the unit responsible for data processing had little influence. No significant disputes between management and unions were reported, but neither did the process include significant information-giving and consultation with the unions. Indeed the rights of trade unions to be informed on this kind of issue were recognised in the Italian system only in the mid-1970s.

During the second phase of development, too, all major steps in the innovation process were implemented as planned by management. There was no divergence between management and unions over the need for change, nor any questioning by the unions of management policies in this area. However, as the possible consequences for employment, work organisation and working conditions began to be realised, the unions started to demand consultation and negotiation, as well as more information. It is only in the third phase – with the widespread installation of terminals – that the unions are beginning to question management decisions regarding the type of technology to be introduced.

2.2.2 The role of negotiation and consultation

During the second phase, which began with the introduction of teleprocessing procedures, disputes and growing divergences were reported throughout banking at both central and peripheral level, concerning the effects of new technologies on employment, working conditions and work organisation. The agreements of 1970 and 1973 consider the eventuality of the new technologies creating new jobs and calling for skills that are not provided for in the existing bargained job classification and wage scale. They provide that the matter will be collectively bargained at enterprise level immediately – even before the reopening of the existing collective agreement. Such was the practice in the bank under scrutiny, as shown by a series of enterprise-level collective agreements in the early 1970s between management and Delegate Council. A *second issue* already settled at national level in 1970 concerned the working time of employees working in EDP (see subsection 2.3.2).

The national industry-wide agreement of 1976 obliges management to inform the local enterprise unions prior to the introduction of new technologies whenever they imply a reduction in total employment. The 1980 national agreement extends this obligation to all cases in which new technologies imply substantial changes in the job content and working conditions of a considerable number of employees. These obligations are part of a wider system of disclosure whereby the enterprise has to inform the trade unions periodically, both at industry and at enterprise level, on all major aspects of its plans for the future. Another issue that has been bargained at national level, and specified in more detail through enterprise agreements, concerns the implications of the new technologies for health, safety and hygiene in the workplace.

The process observed in the bank under study follows the patterns common to the industry, but with some peculiar features. Management have kept the local unions informed of the major steps of technological innovation. On the whole the information has been given prior to the implementation of the major decisions, to some extent going beyond the provisions of the national agreement. Indeed, the high level of unionisation in the bank and the traditional climate of trusting relations between the parties could hardly have allowed a practice of "non-information". Everything is communicated in due time by the right persons, often informally.

This information has usually been given for the whole bank by central management, but specific information has been given concerning individual units, in particular the EDP unit, in these cases directly to the shop delegates. During these information sessions the unions have usually limited themselves to requesting explanations and further details, without objecting or questioning the merit of management decisions. This corresponds to the basic union attitude indicated above of not questioning managerial decisions as such in this area but trying to influence the effects of technological innovations. It also reflects a procedural distinction between information and bargaining sessions. National representatives are not present at these sessions, nor at enterprise-level bargaining. The unions usually handle local issues with a large degree of autonomy (possibly greater in this bank than the industry average).

The innovation process thus went ahead as planned by management; the major disputes in this phase concerned problems of job classification, promotion and consequent remuneration, rest periods and vocational training/retraining of both new and old employees. Some resulted in sectoral strikes. All were settled through negotiations. A national agreement of 1974, for example, indicated the basic criteria to be followed in order to classify (and sometimes to reclassify) employees in the EDP unit, in particular the most skilled. Another national agreement of 1980 entrusted to a *mixed committee* (representatives of banks and unions) the task of examining all problems of job classification arising from new technologies. This committee is very important, and a large number of issues are deferred to it.

An enterprise agreement of 1978, modified in 1985, dealt with health and safety issues, specifying and improving the standards set at national level (see subsection 2.3.2). Vocational training and retraining has been a thorny issue, particularly since 1979, the wide diffusion of terminal teleprocessing having greatly changed the traditional positions of most cashiers and front-office operators. A highly disputed issue concerned the deployment of those employees, many of them middle-aged, who proved not to be "retrainable": a system of job rotation was bargained.

In the third phase, with the more general installation of terminals, the unions are becoming uneasy about the existing practice of information-giving. More information is being requested on increasingly complex matters, particularly on the type of technology the bank intends to introduce, the organisational consequences, the implications for health, safety and working conditions, and the possible impact on job security, which is being felt for the first time.

The unions, including the local unions, seem to have realised that technological innovation has reached a point where the implications are all-pervasive. Pressure from the unions is growing to be consulted over the "whole" reorganisation the bank is undergoing, rather than simply informed. A standing union-management consultative committee was established in late 1985 in order to examine all aspects of the bank restructuring process, including organisational matters. Its activity was, however, uneven and it was suspended after two years. The process of consultation now goes on on a case-by-case, matter-by-matter basis. The power of the unions to bargain over all manpower issues is being used to influence managerial decisions indirectly. On some issues, such as labour mobility, training and retraining, this almost amounts to a power of veto, which obviously maximises the unions' influence on the implementation of all managerial decisions. The decentralisation plan the bank has implemented, for example, which was made possible by the new technologies, was the result of a continuous process of interaction between management and local unions. This process is a mixture of information-giving and consultation over managerial choices and actual bargaining over manpower policies.

Trade unions have recently been involved in a specific aspect of bank reorganisation, namely the "externalisation" of the whole EDP unit. In 1988, the unit was transformed into an autonomous "company" owned in part by the bank itself and in part by other firms belonging to the same group as the bank. This transformation was aimed at improving the efficiency of the EDP service which will be provided by the new company not only to the group firms but also to third parties. The trade unions were informed on all aspects of this transformation and agreed on its beneficial economic effects. They bargained over the conditions of transfer of the EDP personnel to the new company. Indeed, the agreement signed for the new company provides for a system of job classification quite different from the previous one and more suitable to the type of work performed, for a new incentive system, more flexible worktime, and a shift system (see subsection 2.3.2).

2.3 Consequences for the workforce of introducing the new technology

2.3.1 Job security

The problem has not been significant until recently, given the continuous growth of the bank and of employment of all types. Over the years, however, a growing number of employees have proved difficult to retrain and redeploy. In these cases solutions have been found on an individual basis through a sort of informal bargaining between management and employees assisted by the local unions (this being a typical example of day-to-day union work). The most common alternative to transfers has been encouraging employees to retire early, using monetary incentives.

The issue has become more complex in the last few years when the impact of new technologies has reached the point – in this bank as in the whole industry – of blocking the growth in employment and even reversing the trend. Resort to compulsory redundancy or collective dismissals is out of the question – contrary to the whole tradition and philosophy of the bank – and totally unacceptable to the unions. The actual and anticipated reduction in the workforce is relatively small, and partially compensated for by the still good economic growth of the bank. The method most widely used is still encouraging people to retire early. This applies particularly to middle-aged high-level employees and functionaries holding traditional positions of responsibility, who are most directly hit by the new technologies. A supplementary pension plan was agreed upon in 1986 which includes particularly favourable conditions for early retirement. Recruitment has been slowed down but cannot be frozen because new professions and skills are needed.

2.3.2 Work organisation and working conditions

The impact of new technologies on work organisation has been uneven. On the whole the effect has been to broaden the content of jobs rather than to narrow them. The case of terminal operators is among the clearest. While initially the introduction of terminals increased some aspects of "parcellisation" of work, subsequently the process has been reversed. The new positions have joined the traditional tasks of the cashier with some of those of back-office operators. The number of operations that can be done by a terminal operator has increased. This trend towards job enlargement is most evident for employees working in the expanding areas of client consulting, services, etc.

It cannot be said that the unions are playing a major role as regards work design and the planning of procedures and production. The local unions seem less concerned than those in other industries to take part in the planning of the overall work organisation (even though this may ultimately influence working conditions), instead concentrating on ad hoc responses to management initiatives. Management, for their part, have been firmly opposed to any explicit union interference with these decisions, which are considered to fall within their exclusive prerogatives.

The consequences of technological innovation on working conditions, on the other hand, have been dealt with since the early 1970s in both national and local collective agreements. The 1970 national agreement already empowered the banks to organise shift work. It provides for two shifts daily, which may be fixed between 6 a.m. and 10.30 p.m. A third shift may be introduced only for particular technical reasons and within maximum individual limits of 80 days a year (female employees are excluded), within time-limits to be bargained with the local unions. A rest period of 20 minutes is granted to employees working on a continuous shift of 7 hours 45 minutes, and a rest period of 30 minutes for employees working on punching machines. These provisions were reproduced in all subsequent agreements until the last one of 1987, which increased indemnities due to employees working shifts.

Connected to these are the provisions for rest periods for employees working *exclusively* at video-terminals – not those using video-terminals as an instrument for their work. The national agreement of 1987 prescribes a pause of 15 minutes every two hours. (Further reductions are provided for employees working shifts.) Enterprises are committed to employing in other tasks employees who are recognised to be unfit for work at video-terminals, according to medical checks by specialised public institutions. These national provisions were anticipated in the bank under study by a collective agreement of 1985.

Safety and health are recurrent issues in national and local bargaining. An enterprise agreement in 1978 already contained a general statement of policy committing the bank to implementing all measures apt to reduce the negative consequences (noise, temperature, dust) of the introduction of new machinery and to improve the working environment (air conditioning). The same agreement entitled employees to ask for periodical medical checks. Although the agreement did not refer specifically to employees working with new technology, the provision is made most frequent use of by this group.

The unions also have the right to check the working conditions and environment and to ask the advice, at the firm's expense, of experts from specialised public bodies. This provision has been used quite often, on average two or three times a year. In addition, the unions frequently request the management informally to put into effect specific changes in the working environment directed to improving working conditions.

2.3.3 *Payment systems and income protection*

The influence of new technologies on wage scales is only indirect, that is, through the changes introduced in work organisation and in job contents. As indicated above, the classification of job positions resulting from technological innovation has been one of the major objects of enterprise bargaining since the mid-1970s. In 1980 it was entrusted to a special joint committee. On the whole, the impact of technological innovations has not greatly altered the wage levels and relative positions of the workforce. This is due to the fact that wage scales are fairly static and basically fixed by the national agreement. Personal ad hoc allowances are not admitted by the unions and wages linked directly to productivity do not exist. The so-called production bonus bargained yearly is distributed to employees in proportion to their position in the classification system.

A significant case is that of employees working in the area of EDP. The emergence of new skills in this area has been recognised since 1974 by an agreement granting privileged treatment to these employees. Their peculiar position was confirmed by the new system of classification negotiated in 1988 following the transformation of the EDP unit into an autonomous enterprise. The five traditional job positions were "spread" between seven new job levels. This increased relative wage differentials among the various groups of employees. Automatic advancement has been reduced and replaced by a more discretionary process.

The possibility of downgrading employees, in the sense of reducing their position in the job classification scale and hence their wages, was in any case merely theoretical, given the bank's tradition of labour relations and the unions' power. In most cases such an outright downgrading would have been against the law (Act No. 303/1970, article 13).

The most difficult issue in this respect has been finding a redeployment that is acceptable to the interested parties, particularly when the decision implies a change of place of work: geographical mobility is very hard to achieve, and has to be reached through a lengthy process of informal bargaining and "lobbying".

2.3.4 Training and retraining

Training and retraining have acquired growing importance in recent years. It is generally felt by both parties to be a necessary instrument to adjust personnel, at all levels, to technological change. According to the 1980 agreement, the bank is committed to providing all workers affected by technological change with the training they need during regular working time.

Priorities and exceptions, including cases of employees refusing to accept technological change, are evaluated on an individual basis, through informal bargaining. Usually the formulation of training programmes is left to management. The unions are kept informed of the type of courses to be implemented. Recently labour-management cooperation over training and retraining has been increasing. This is particularly the case for the training of new young employees, who are being hired on a special fixed-term contract (24 months) the so-called education contract, which provides for part-time work and part-time training. In fact this type of contract receives monetary incentives under Italian law and from the European Social Fund. Union agreement on the training programme is required both by the European regulations and by Italian law. In March 1987 banks and unions agreed on a programme to train a considerable number of newly hired young employees in the use of new technologies: 20 full days out of a 43-day total training period.

The matter of training and retraining has been considered by the standing consultative committee established in late 1985 to examine jointly all aspects of the bank's restructuring. Lately it has been discussed as such in ad hoc meetings.

2.4 Effects of the new technology on labour relations

2.4.1 Effects on the structure of the workforce

As indicated above, the new technologies have not had a significant impact on the size of the workforce, but the number of workers employed in the areas affected by the new technologies has been growing rapidly. Out of 3,300 employees in early 1987, 141 are specialists employed in the central EDP, while over 600 work exclusively at video terminals. This number has been growing particularly in the third phase of development.

More favourable career opportunities have emerged only for the employees of the central EDP. Those employees in peripheral terminals remain in the traditional career patterns (in part based on seniority). Initially, the bank resorted mainly to external specialists for EPP, but labour mobility is now increasing. Employees are regularly retrained from traditional jobs in order to be employed at peripheral terminals and even in central EDP. In 1985, for example, an internal competition was organised to select ten EDP specialists and over 100 offers were submitted from various sectors of the bank. As a policy the bank prefers internal promotion to external recruiting as far as possible.

Geographical mobility has also increased as a consequence of technological change and of the programme of structural decentralisation of the bank.

2.4.2 Effects on the unions

No significant change in union membership, structure or militancy is noticeable over the last few years. Union membership in EDP does not differ from the bank average. No new association of "professional" employees has been formed in competition with the traditional unions, as has occurred in other banks. The existing unions have been prompt in representing all employees, including the newly hired specialists. This confirms the strong "job" consciousness and "business" attitude of these unions in the banking industry.

On the whole the national unions have played a positive role in the process. Sometimes they have intervened to smooth the resistance of local groups and union members to specific changes.

2.4.3 Effects on management

The impact of technological change on management seems more pronounced. While the decisions pertaining to innovation have mainly been centralised, the use of terminals and of diffuse information systems, particularly in the third phase, has brought about a wide decentralisation of services and organisation.

Management functions related to organisation and planning have acquired greater importance, due to the need to design new organisation patterns and adapt the work environment to technological change. Personnel and human resources management played a relatively secondary role during the first two phases. Their main tasks was to fit the new specialised EDP jobs into the rather rigid job classification system common to the banking industry. The role has been growing recently, given the need to revise a larger number of job profiles, to promote greater labour mobility through concerted massive training, and to inform and consult with the unions on the innovation process. The personnel function is itself being in part decentralised in order to match the decentralisation process taking place in the bank as a whole.

No significant change is reported in the relationship between the management of the bank and the employers' association. The bank follows the association directives in national bargaining but remains by and large autonomous in the introduction of technological change and subsequent manpower policies.

2.4.4 *Patterns of negotiation and consultation*

The trend in recent years has been towards a more cooperative pattern of negotiation and consultation, with a significant reduction in conflict. Union-management contacts have always been fairly frequent, according to the bank tradition. This trend has been confirmed and stabilised in recent years, together with a more systematic use of information and consultation in lieu of adversarial bargaining.

2.4.5 *Conflict*

The reduction in conflict corresponds, however, to a general trend in Italian labour relations which is due to more complex factors than technological change. But new forms of strike have sometimes been adopted in connection with the type of technology. The employees working in peripheral branches using mini-computers have gone on strike at different times from other branches connected to or dependent on the same computers in order to maximise the impact of the strike on the bank and minimise the loss by the employees. In other cases, however, the very technology has allowed the bank to minimise the damaging effects of strike action, for example, by the employees working on central EDP machines.

2.5 *Evaluation*

The main characteristic of labour relations in this bank all contribute to maintaining stability: the tradition of management, the strength of the local unions, and their pragmatic and firm-oriented attitude. These same characteristics account for the fairly smooth transition from the traditional pattern of organisation to the diffused use of new information technologies. The possible negative effects of technological innovation have been reduced to a minimum. Employees have been by and large protected in relation to wages, working conditions, job security and (mostly) professional status, with exceptions limited to some older workers mostly middle to high-level functionaries.

Union-management relations prevailing in the bank have contributed to resolving the problems arising in the process of technological change in a satisfactory manner with a low level of conflict, as had been the bank tradition in previous more stable periods. The growth of bank services and profits has certainly made this task easier.

The degree of mutual satisfaction of the parties may vary. Recently some of the more militant union groups have complained that their involvement in the process is limited to receiving selected information and to discussing *ex post* the *effects* of new technologies while being excluded from the decisive phase of planning and choosing between basic options. But these complaints are hardly likely to alter the main trends, unless a worsening of the economic prospects of the bank threatens to prejudice the present (satisfactory) state of working conditions and job security – which is unlikely.

This pattern of union-management interaction, which may be termed "integrative", or of "limited" union involvement, is common in Italian labour relations. Moreover, a basic distinction is maintained, although informally, between different matters. Management control usually remains unchallenged, although allowing for information and consultation with the unions, in such areas as investment and strategic decisions – including technological innovation; and this is true even in those enterprises most inclined to cooperative relations with the unions (such as the public groups, the Institute of Industrial Reconstruction/Istituto Ricostruzione Industriale – IRI – and the National Hydrocarbon Corporation/Ente Nazionale Idrocarburi – ENI). Collective bargaining remains the main channel for determining wages and working conditions while some degree of employees' control over such matters as hiring, careers and work organisation exists in some cases of strong business unionism.

4 Japan

Technological change and labour relations in Japan

*Yasuo Kuwahara**

1. Introduction

Technological innovations involving microelectronics have diffused in various areas of industry and daily life in Japan since the latter half of the 1970s at a speed and scale beyond initial expectations. In particular, the role played by labour and management as this occurs has become an issue. A new technology can spread throughout the whole of society if labour and management respond flexibly at the grass-roots level, allowing the technology to be introduced smoothly. This leads to modernisation of the industry and a strengthening of competitive power.

In general, both labour and management in modern Japan are positive and cooperative about technological innovation. One reason for this is the accumulation of experience in the post-Second World War period. The intense disputes between labour and management during the rationalisation process that accompanied the technological innovations in the 1950s and 1960s in industries such as mining, cars, chemicals, and paper and pulp often caused deep-seated, lasting resentments between labour and management. Neither labour nor management want this to happen again. A second reason is that competition between enterprises has intensified. Trying to hold back new technology would lead to a weakening of an enterprise's ability to compete. Thirdly, as long as the idea of lifetime employment is widespread in Japan, the success or failure of an enterprise must greatly affect the lives of its employees. Innovative firms have advantages over unprogressive ones, including attractiveness to workers.

The key participants in the process of responding to technological innovations are labour and management, at enterprise level; the roles played by government or other regulations are relatively small. Political parties and national union organisations do comment on the importance of the new microelectronics technology, but their influence is small compared to that of the grass-roots response from labour and management.

This case-study will provide many, hopefully valuable, pointers towards understanding the actual state of diffusion of new technology in Japan. While there have been many studies on the general response from labour and management to the introduction of new technology,[1] detailed case-studies are relatively few in number.[2] One possible reason for this is that there are few cases of labour-management disputes over the introduction of new technology.

* Dokkyo University, Japan.

One general finding is that labour savings often occur as a result of the introduction of microelectronics technology – as would be expected with machinery such as industrial robots and numerically controlled (NC) machine tools whose very purpose is to save labour. Nevertheless, labour savings are not always realised, and in some cases the number of employees actually increased after the new technology was introduced.

This so-called "jobless growth" or "disemployment" has been taking place in manufacturing industry since the 1970s.[3] Employment in manufacturing has recovered somewhat in the 1980s, but the rate of growth is significantly lower than in the period up to the 1960s. Meanwhile, growth in employment in service or tertiary industry has been consistently high, and was so even during the depression of the 1970s.

In addition, the number of production workers in manufacturing industry decreased significantly during this process. This was offset by a steady growth in the number of white-collar workers in the broad sense of the term, including those engaged in office and service work. The number of blue-collar workers in traditional areas of industry such as steel, heavy machinery and shipbuilding has decreased, while the number of service and software-related workers has increased.

Traditionally, the major industrial *locus* of labour unions in Japan is manufacturing industry, particularly sectors such as steel, metals, chemicals and machinery.[4] Union density in service industry has always been low. Until the low-growth period of the mid-1970s, when growth in manufacturing industry on the whole became sluggish both in production volume and in employment, the number of union members was expected to increase in proportion to the expansion of these industries, given that most labour unions in Japan had adopted the union shop clause.

Service industry, on the other hand, was expanding, but this led to an overall decrease in union density. Unionisation dropped continuously up to the mid-1980s, reaching 28 per cent by 1986. The future strategy of labour unions should therefore be to increase their membership in poorly organised sectors such as service industry. One reason for their lack of success so far is that service industry has so many small establishments and is so much more decentralised than manufacturing industry.

The following case-studies look at the impact of technological change in a printing firm, a machine-tool manufacturer and a bank. The latter should be particularly valuable, as information on the impact of technological innovation on service industry is relatively scarce.

2. Case study of a printing firm

2.1 The context

2.1.1 The printing industry

Before we start we should define exactly what is meant by "the printing industry" – basically those firms included in the Japan Standard Industry Classification SIC-25 ("publishing, printing and relation industries"), minus SIC-251 ("newspaper industry") and SIC-252 ("publishing industry").

Two basic industrial characteristics of the printing industry in Japan are worth noting at the outset. First, with the exception of a few large corporations, printing firms are predominantly small in scale, and division of labour in the local markets is the norm.[5] It has proved easy to develop a system of specialisation between the various firms, and unstable demand in the printing industry has not favoured an increase in the size of firms. In addition, the industry is heavily concentrated in cities, where demand for printing is high.

Second, since printing involves a large number and great variety of small production runs, standardisation is difficult and manual skills are still necessary to a great extent. As a result, labour costs account for an extremely high proportion of production costs, even though printing is often regarded as a very profitable industry. Furthermore, despite gradual improvements, until recently printing technology was basically unchanged since the Gutenberg days of picking out metal type to make a plate.

It was not until the late 1960s that the traditional hot-metal composition process began to be replaced by new technology for plate-making called "cold-type" or CRT (cathode ray tube) composition, such as phototypesetting and computer typesetting. Another important technological change was the development of the colour scanner, fully equipped with the latest electronics. The process of manual colour separation, using a plate-making camera, which had required great skill and much time, was thus substantially automated. The traditional production process has thus been transformed by the development of microelectronics technology.

2.1.2 The enterprise

Company X is a general printing and paperware company, one of the few large, prominent printing firms in the Japanese printing industry. Although the workforce has decreased in number somewhat in recent years, there are nearly 9,000 employees. The firm's capital was approximately 40 billion yen as of early 1987. It can truly be regarded as an "elephant" in an industry that has often been referred to as consisting of "elephants and ants". Although, because of its size, it is by no means typical of printing firms in Japan, Company X was chosen for the case-study because it has led the industry in technological innovation, management innovation, labour-management relations, and other areas.

2.1.3 Labour relations prior to the introduction of the new technology

The printing industry has from pre-war days been one of the mainstays of the labour movement; one of its earliest gains was the establishment in 1919 of the eight-hour working day by the Japanese Printers' Union (Shinyu-kai). The union of Company X had had years of turbulent industrial disputes by the 1950s. However, it retired from the National Federation of Printing and Publishing Industry Workers' Unions (Zen'insoren) largely because of differences in policy direction. Since then, it has maintained a cooperative relationship with the employers.

As in most Japanese firms, the union is an enterprise union. It has its headquarters at the firm's head office and branches at its various establishments. There is a branch manager in charge at each establishment, with three key officials, including the chairman and chief secretary, who are elected by the branches, in charge at the headquarters. Depending on the nature of the issue, union-management negotiations are carried out at either establishment or head-office level.

The main instrument for union-management consultation at head-office level is the Management Council, consisting of representatives of labour and management. Usually, the Management Council meets once a month; discussions are mostly about operations-related issues such as orders and production plans. However, issues related to the introduction of new technology are also submitted there. In fact, the labour agreement states that items related to (1) management policy and management conditions, (2) setting up new establishments, or the merger or bankruptcy of the firm, and (3) introduction of major new facilities or technology, should be submitted to the Management Council.

2.1.4 Reasons for introducing the new technology

Around the mid-1970s Company X decided to take the plunge and introduce the first plate-making facilities. This was the beginning of the various technological innovations that the firm has been pursuing until the present day. At that time, the Japanese economy had undergone a period of high growth, Company X had become the leading firm in the industry, and demand for printed matter was expanding rapidly.

With the increasing demand, manpower shortages had become a serious problem. Printing firms were having great difficulty in recruiting young workers because of the negative image persisting since pre-war days of dark, dingy workplaces. Furthermore, the technical processes of printing at the time demanded a high level of skill on the part of typesetters and printers, which required much experience and training. Introducing new technology seemed to be a way both of solving these recruitment problems and of enabling the firm to expand production to meet demand.

The fact that labour disputes, which used to occur frequently, had become very rare was also very much to the advantage of Company X during the large-scale technological innovation that began in the 1970s.

2.2 The decision-making process

2.2.1 Decision-making regarding the introduction of the new technology

Both the decision to introduce new technology and the choice of technology to be introduced are usually the prerogative of management in almost all industries. However, the introduction of new technology is in principle submitted to labour-management consultation from the planning stage. Since Company X is a general enterprise and its market involves a wide range of areas in printing (including electronic PC boards and publishing), the facilities introduced are quite diverse, ranging from personal computers to colour scanners costing several hundred million yen per unit. The introduction of all such facilities is normally discussed through the labour-management consultation system established by labour agreement. As already seen, the labour agreement stipulates that management will submit all plans to introduce new technology to the Management Council. There is an understanding between labour and management that issues related to the introduction of new technology can be resolved within the framework of the conventional labour agreement, so no separate agreement has been concluded to deal with new technology. If agreement cannot be reached in these consultations, collective bargaining is carried out. In most cases, however, the issues are resolved within the Management Council.

2.2.2 *The role of negotiation and consultation*

If personnel redeployment is necessary because of the introduction of new facilities, the previously mentioned labour-management consultation system based on the labour agreement is activated. That is, regardless of whether the redeployment involves only one establishment or several (or is even confined to one workplace), an initial discussion is held at head-office level by the Management Council. Collective bargaining follows if an understanding between labour and management cannot be reached at this stage, but this very rarely happens.

If the union has agreed, the Council also discusses the methods of training in the skills necessary to operate the new facilities. If necessary, workers will be dispatched to the equipment manufacturer to acquire the skills. The decision on whether or not technical training is necessary is determined by consultation between the firm and technical experts supplied by the union.

Normally, the Management Council comes to the conclusion that the issue will be better discussed at establishment level, where the ramifications can be more easily grasped. The Central Executive Committee of the union, comprised of branch managers, issues statements to the respective branches. After discussing the contents in detail, each branch holds discussions with management representatives at the establishment – this is known as the "Labour-Management Meeting".

Depending on the issue, subcommittees comprised of union members who are informed about the topic may hold discussions at branch level. The results are studied by the branch committee of the union and the firm's management, and a report is sent to higher officials.

Since issues such as labour-saving and changes in personnel have a significant impact on the workplace, they require the opinions of people who are on the spot. A subcommittee comprised of members elected from the workplace and the person in charge of the workplace is therefore formed to discuss the content of the management suggestion. This derives from the basic understanding between labour and management that the workplace is the basic unit. A final decision on a report from this meeting is made either at head-office or at establishment level.

2.3 Consequences for the workforce of introducing the new technology

2.3.1 Job security

A study on changes in the size of the workforce shows that it reached a peak in 1974 and had dropped to 15 per cent below that level by 1982. Meanwhile, sales grew by 155 per cent during this period.

The production division dealt with the streamlining process that accompanied the introduction of the new technology through attrition. In other words, employees who retired were not replaced (the current age of retirement is 60) and recruitment was reduced. Meanwhile, the sales division slowly increased in size in order to cope with the intensified competition; it is in any case an area in which labour-saving is intrinsically difficult.

Since Company X's establishments operate a three-shift system, including a night shift, the percentage of female employees is low. Since female employees generally leave a firm sooner than male workers, considerable elasticity can be anticipated for personnel planning in industries in which the percentage of women is high. This is not the case in Company X, so management took a very cautious approach to personnel planning.

Up to now, the management have avoided redundancies almost entirely, although they have never ruled out the possibility. This approach is possible for Company X because it is basically one of the leading firms in the industry, with excellent research and development capacity, and increasing employment opportunities.

Redeployment was also kept to a minimum, due to the firm's policy of introducing the new technology gradually, placing new facilities side by side with old facilities, so that no jobs were completely displaced. As a result there were very few cases of personnel relocation where a worker was reluctant to be transferred to the new facility.

2.3.2 Work organisation and working conditions

Given the firm's policy of introducing the new technology gradually, the old printing skills still survive alongside the new ones. Anyone visiting the Tokyo establishment of this firm would be amazed at the contrast between, on the one hand, middle-aged workers picking out type with tweezers and, on the other, workers dressed in white operating new facilities that are said to be the most advanced in the world, both within the same establishment. The scene is almost like a museum history of printing technology. It is especially amazing in that there are said no longer to be any medium-sized or large firms in Tokyo that use the hot-metal process.

The basic skills required in the printing industry used to take many years to master. Jobs such as type-picking and typesetting, for instance, required great skill: as well as being physically hard work because of the heavy lead types, a high level of precision was required. It was said that three to seven years of training were necessary for a worker to become fully qualified. Thanks to the technological revolution, however, it is possible for a person to be able to use photocomposing machines after a month of training and to become fully qualified after six months. Moreover, it is no longer hard physical labour, which makes it possible to recruit women workers – though in the case of Company X, the typesetters are still all men, due to the three-shift system.

Colour separation has also been revolutionised by the development of the colour scanner. It has been said that ten years were needed before a worker could become proficient in visualising the colour net, making quality control a big problem because it dependend largely on intuition and experience. With the introduction of the scanner, however, the time was reduced to six months and standardisation of quality became possible. In the case of Company X, however, work using cameras is still performed despite the introduction of the colour scanner, so operators of traditional machines and retouch-men are still to be found.

There can be no doubt that the importance of traditional skills will gradually decline, and increasing numbers of workers will be machine operators. On the other hand, it is said that retouch-men will always be necessary. In the past, the skill of the retouch-man was said to determine the quality of the offset plate. Even today, when the colour scanner has become the norm, their ability and sensitivity is still needed. The skills of the retouch-man in relation to graphics, aesthetics, colour and other matters are a vital factor in producing high-quality work; a knowledge of colour separation and the whole printing process is indispensable.

In addition, since printing work requires a high level of interaction between the skills, there is a limit to how much the human factor can be removed, and workers' experience is still important. That is why length of service is, in general, still regarded as significant.

2.3.3 *Payment systems*

In the decade after the oil crisis, with an ageing workforce, management made several attempts to change the system according to which wages increased in proportion to length of service, because it was naturally leading to increased labour costs. It then became policy to guarantee employment until retirement age, while halting the increase in wages at, say, 45 years old.

Moreover, younger workers were gradually increasing in number in the union, leading to a shift in the power structure, and they started demanding higher wages and bonuses for their age group. These changes in the remuneration system have on the whole worked to the disadvantage of middle-aged and older workers.

2.3.4 *Training and retraining*

Acquisition of skills in the printing industry used generally to happen through on-the-job training. A worker posted at a particular section acquired skills by observing or receiving guidance from an experienced person. This method could be used and became widespread because the technology did not change fundamentally over a long period.

However, the rapid progress of technological innovation since the latter half of the 1970s has significantly altered the old system. In the case of Company X, workers with many years of service were kept posted at old facilities, while new workers who had just graduated from junior high school or high school were posted at new facilities. In most cases, the skills necessary for operating the new facilities were taught by engineers dispatched from the equipment manufacturers. Alternatively, employees of Company X were dispatched to the manufacturers for a certain period of time to acquire the skills they needed. Labour-management consultations took place on all such changes in the methods of training and retraining. Issues related to the ageing of employees have been increasingly talked about at labour-management consultations in recent years. Nothing can be done about the decrease in learning ability of older employees, and the new technology requires more and more education and training. Solving this problem has become one of the major tasks for labour and management.

2.4 *Effects of the new technology on labour relations*

2.4.1 *Effects on the structure of the workforce*

As far as employment sturcture is concerned, the number of office workers and sales staff increased slightly, while the number of production workers declined significantly. It is skilled workers engaged in traditional hot-metal printing, particularly middle-aged and older workers, who have felt the impact of the new technology the most. Although there were no redundancies at Company X, such workers were put at a relative disadvantage in terms of their future careers.

The fact that the time required to acquire the new skills is so much less than that for the traditional skills is thought to have increased both the intra-firm mobility of workers and their inter-firm mobility. A high proportion of workers at Company X will never leave the firm for another job, but the mobility of workers between smaller enterprises is expected to increase in the future.

2.4.2 *Effects on the unions*

The basic approach of the labour union to technological innovation is a positive one. Their belief is that technological innovation is inevitable in order to survive the competition between firms in the free-market economy. The union therefore makes efforts to have the management inform it about any new facilities and anticipated changes in the workplace at the earliest possible time in order to strengthen the system of prior consultation.

2.4.3 *Effects on management*

Technological developments have had more impact on management than on the union. In particular, the improvements in productivity and quality brought about by the new technology, as well as the impact on corporate revenue, have become important pillars of management strategy. This is not limited to the printing industry, and seems to be one of the reasons for the increase in managers with technical backgrounds in large enterprises since the 1970s.

2.4.4 *Patterns of negotiation and consultation*

One effect of the new technology has been an increasing decentralisation of negotiation and consultation. Company X's establishments possess diverse technological systems, and the introduction of new technology largely depends on the particular situation at each workplace. Negotiations and consultations between labour and management about the introduction of new technology therefore filter gradually down to the establishment or workplace level. As long as no serious problems exist, the actual decisions are made at those levels.

2.4.5 *Conflict*

The introduction of microelectronics technology in Company X has so far progressed smoothly without the intense labour-management confrontations that took place in the past. One possible reason for this is that both labour and management had learned their lessons from the bitter experiences of the early disputes, and wanted the introduction of the new technology to be peaceful. In addition, the intensified competition in the market is likely to have had a dampening effect in the long term on labour disputes. Both management and employees of Company X have a tremendous interest in market trends, and in the working conditions of other major enterprises.

Furthermore, as mentioned previously, a principle of Company X is to establish new facilities side by side with the existing ones, rather than replacing the old with the new immediately, thus avoiding the need for extensive redeployment.

Lastly, being one of the largest firms in the industry, employment opportunities in Company X were increasing, given its consistently favourable business results during the high-growth period in the 1960s and the rapid expansion of new fields in the printing industry. Employees were therefore able to be certain of employment and financial security as long as they were willing, if necessary, to be redeployed to different types of job, and the issue of large-scale redundancies never arose. In smaller firms, on the other hand, where business is generally confined to a fairly narrow range of activities, there is little scope for redeployment and redundancies were inevitable in the streamlining process following the introduction of new technology. Many of these firms became involved in very intense disputes.

3. Case-study of a bank

3.1 *The context*

3.1.1 *The banking industry*

In Japan, banking is one of the industries, like insurance and general trading, in which computerisation is most advanced. Computers and the on-line systems based on them are the nerve centres of office automation in banking. Considering the number of transactions that must be processed, and the very wide range of business covered by the systems, banking can be said to be the most advanced industry from the point of view of office automation.

The development of office automation in the Japanese banking industry can be divided into several historical phases. The first is the period from 1945 to 1960, when items of equipment such as typewriters, calculators and banknote counters were introduced, mainly as back-up for manual operations. Until this period, abacuses and telephones were the only items of equipment used in banks. Around 1955, punched-card systems were introduced, and business was standardised. This formed the basis for the computerisation that followed.

The second phase lasted roughly from 1960 to 1965. In this period, banks were popularised, and the number of transactions increased greatly. Accordingly, the number of employees was increased and computers were introduced to deal with the greater workload. Clerical work and over-the-counter work came to be handled mostly by women. With computerisation, each bank organised central data processing on a batch system, but there were delays because the system was not on-line.

The third phase lasted from about 1965 to 1975, during which first-stage on-line systems were developed. This was facilitated by cheaper and better computers and improvements in technology, such as in telecommunications systems. Automated, on-line processing systems for ordinary deposits were introduced. Ledgers at individual offices became unnecessary, and business was greatly rationalised. Although there was labour-saving generally, the number of staff working with computers increased. Customers benefited from the improvements in services: for example, withdrawal was made possible at any branch.

First-stage on-line systems processed different types of business – deposits, exchange business, etc. – separately. The expansion of services was therefore limited in that a new system had to be created for each new service. The fourth phase covers the period from 1975 until the present day, during which second-stage on-line systems were developed, the main characteristics of which is the capability for on-line processing of all branches and types of business. This system has brought benefits such as better organised customer files and easy retrieval of files by customer name and address. It is supported by the many types of automated terminals which have been introduced, including cash dispensers and automated telling machines (ATMs). Such operations as transferral, paying-in/withdrawal and bill payment can now be automatically processed, as can entries in the ledger, summaries of accounts and daily trial balances. All this has resulted in dramatic labour-saving and rationalisation.

Inter-bank on-line systems also exist, making business tie-ups and processing between banks possible. It is therefore becoming increasingly difficult for a bank to choose new equipment independently. Moreover, if a bank does not install certain new facilities, it will become uncompetitive. As a result, labour unions cannot gainsay the introduction of new technology unless it is clearly disadvantageous for workers. In general, the unions are fairly cooperative regarding the introduction of new technology.

The range of such systems is still developing and expanding, but it is expected that they will spread beyond bank branches to cover customers' houses and offices, leading to third-stage on-line systems. It is thought that city banks will introduce such systems within two or three years. The level of office automation in Japanese city banks is thought to be the most advanced in the world.

3.1.2 Labour relations in the banking industry

Labour relations in banking are typical of industries employing mainly white-collar workers. Unions are enterprise unions, and these affiliate to the Japan Federation of Bank Employees' Unions (Zenginsoren), formed in 1957. The Japan Federation works during the Spring Offensive (annual wage drive) and at other times to improve working conditions at industry level by presenting a unified claim for the whole industry. However, enterprise unions play a decisive role at enterprise level.

White-collar worker labour unions, such as bank employees' unions, have traditionally been progressive and radical, contributing greatly to the modernisation of Japan after the war. At present, however, their industry-level activities are rather conservative compared to those of labour unions in other industries.

The Ministry of Finance also exercises considerable control over banking in order to ensure that excessive competition is avoided. Changes in the number of branches, the deposit rate, etc., are therefore severely restricted and individual banks have to operate within these limits.

3.1.3 The enterprise

Bank Y, the subject of this study, is one of the 13 banks classified as city banks, and is ranked high in this category. The bank has about 200 branches, and is known to be a leader in terms of modernisation.

Labour savings became possible at each branch through the introduction of machines operated directly by customers. At present, 70 per cent of withdrawals, for example, are handled by customers themselves, which has contributed greatly to labour savings. With the introduction of machines and on-line systems, the total number of deposit accounts has reached 20 million, and about 500,000 people visit the bank every day. Because of the new technology, the bank can deal with the increased volume of transactions without increasing the number of employees.

3.1.4 General characteristics of employment relationships

Banking is typical white-collar worker employment. Since the Equal Employment Opportunity Law was enacted in 1986, new trends in employee distribution may be emerging, but traditionally there have been clear demarcations according to educational background and sex: those employees expected to fill managerial positions in the future are male university graduates; core clerical personnel are mainly male high-school graduates; and lower-level office employees are mainly female high-school graduates.

Bank Y uses an employment rotation system (a similar system is used by other city and regional banks), according to which most employees (except directors) are rotated every three to four years. Usually, new university graduates work in the back office for one year without any direct contact with customers. From the second year, they are moved to the loaning and sales fields (mostly in local branches) in order to gain experience of direct contact with customers. After this they are promoted in the bank by being rotated or transferred every three to four years. Male high-school graduates are rotated and promoted in a similar way. The system was devised to ensure that bank workers in these categories have a wide range of job experience throughout their careers. Female high-school graduates used to work almost entirely in the local office, and for them transfer between branches was very exceptional. Distinctions like this may disappear due to the law regarding equal employment opportunities for men and women.

3.2 The decision-making process

3.2.1 Decision-making regarding the introduction of the new technology

In Bank Y, new technology is studied and its introduction planned by the office work division at the head office, taking account of both internal and external trends. The office work division then submits the basic plan to a meeting of the managing directors, the top decision-making body, for directions, and subsequently reports on progress when required.

Business at branch offices where new facilities are introduced can be broadly divided into three areas for consideration: over-the-counter business such as deposits and withdrawals; loan and other sales-related business with customers; and back-office business which supports the other two areas. The office work division analyses activities in these three areas, and tries to realise rationalisation, automation and labour savings.

3.2.2 The role of negotiation and consultation

In Bank Y, a labour-management meeting is held once a month at head-office level. At these meetings, there are explanations and consultations about new technology and rationalisation measures long before they are actually introduced. Consultations about the on-line system, in particular, started very early because it would affect all branches so significantly, and employees were kept well informed through both management and union.

The nature of labour-management consultations varies according to the subject. Rationalisation plans are usually discussed in great detail because the introduction of new technology affects all workers and cannot work at all without their complete cooperation. Consultations cover the overall schedule for the introduction of new equipment, the daily progress schedule, the specifications and operating system of the equipment, effects on the workforce, and so on. Items specific to a particular branch or workplace are discussed at that level by representatives from the union branch and management.

3.3 Consequences for the workforce of introducing the new technology

3.3.1 Job security

Partly because of the various restraints imposed by the Ministry of Finance – for example, opening, reorganising or closing a branch needs approval from the Ministry – and partly because of the increase in the overall volume of bank work, there have been no redundancies due to technological innovations. During the 1960s, business expanded sharply with the introduction of primary on-line systems at a time when economic growth was in general very rapid.

From around 1978, however, some banks did begin to reduce the number of employees. This was partly because in 1977 the Ministry of Finance made it easier to obtain approval for setting up small-sized branches (with 15 or fewer workers) and automated branches (with five or fewer workers). In fact one branch of the bank has a staff of only four; it is now possible to increase the number of branches without increasing the total number of employees.

Office automation systems have greatly affected the need for labour, particularly in offices that handle customers directly. In the past, Bank Y had 70-80 employees in medium-sized branches, but the number is now less than 50, including salespersons for individual customers. Out of this workforce, 20-30 people are involved in clerical work. The number of workers of all types is decreasing, but particularly the number of female workers. In city banks, recruitment of female employees has decreased by 20-50 per cent since 1983. In the past, double-checking was a strict requirement in banking, and this hindered labour savings. Once office automation equipment was introduced, however, these tasks were taken over by machine, leading to a sharp reduction in the back-office workload. Rationalisation through the further use of machines for what are now manual activities will continue for some time.

Although it is not spelt out by union agreement, it is tacitly understood that regular employees are never dismissed before retirement except in extraordinary circumstances.

Each bank is therefore very careful regarding personnel planning and management, and reductions in the number of employees are handled according to the "attrition principle". This policy has two arms: temporarily transferring or dispatching middle-aged and older staff to affiliated companies, and reducing the recruitment of new graduates.

The surplus staff made available by the automation of work are being shifted to areas such as direct sales to customers, financial affairs and fund operations. Banks are today perceived as providing favourable employment opportunities in terms of job environment and remuneration. Bank Y recruits 200 new university graduates (mostly male), 100 male high-school graduates and 200 female high-school graduates each year.

3.3.2 Work organisation and working conditions

The general outcome of automation is that the content of work for most employees has narrowed, and what is left for human labour tends to be of a higher grade, such as business consultation or financial planning.

As far as working conditions are concerned, there have been minor problems. At the computer centre, for example, the introduction of the three-shift system and the use of temporary workers from an outside computer firm caused some unrest. Fatigue related to the use of video-terminals and general health problems in the new working environment were frequent subjects for negotiation.

Previously, shift-work in banks had been limited to security guards or gatekeepers. With the development of office automation, however, the volume of work at the data-processing centre increased greatly, necessitating other night workers for data processing and the maintenance of machinery. Out of about 200 computer operators, 80 work a three-shift system. Workers in the computer or development divisions also have different occupational classifications from the conventional bank workers. Although none of these issues is thought to be serious at the present time, the union holds discussions on them on the assumption that they may become important in the future.

3.3.3 Training and retraining

On-the-job training is the basic method of training for bank workers, with some additional off-the-job training. As already seen, many employees are rotated every three or four years, and thus receive training for the new jobs using the new equipment. New employees are given practical training in the operation of basic office machines at the time they join the bank. All university graduates without exception work for some time at a local branch, and are given experience in jobs such as teller which involve direct contact with customers.

3.4 Effects of the new technology on labour relations

3.4.1 Effects on the structure of the workforce

Technological innovations have caused many changes in banking. As already seen, the number of workers per branch has certainly decreased from the peak in 1976. The introduction of computers necessitated an increase in staff for electronic data processing (EDP), but the number of female workers decreased as clerical work was rationalised and simplified.

Another new trend that has emerged to cope with the changing work volume is the employment of part-time workers. Female ex-employees who stopped working when they got married or for some other reason are often re-employed in this capacity, as they have a good knowledge of the work and do no need any in-bank training. There are some advantages for the bank in having part-time employees: for example, bonuses and fixed expenses are less than for regular employees. For their part, these part-time workers like returning to the place where they used to work. In order to maintain relationships with female ex-employees, major city banks issue newsletters telling of employment opportunities and other relevant news.

It seems almost certain that these trends will continue, with further decreases in the number of employees per branch, and a further shift away from simple clerical work in the back office to higher-level work requiring discernment in the fields of finance, fund operation and consulting. Because of the increasingly severe market competition, sales efforts to customers will have to be heightened, and the amount of work related to EDP will increase further and will be undertaken most by part-time workers.

3.4.2 Effects on the unions

Because of the development of on-line systems since the 1960s and the Ministry of Finance's recent policy of deregulation, competition and differentiation between banks has increased. Other general trends include a greater commitment on the part of employees to the bank they work for, and a greater willingness on the part of employees to become union officers during their careers, since it is now recognised that this experience is not in any way disadvantageous later when an employee reaches managerial level and loses union membership status. In Japanese enterprise-based unions, full-time union officers appointed by the union members can usually retain their status as employees of the firm, and quite often become ordinary workers again when their term of office expires. In banking, in fact, there are many directors who were once labour union leaders. A president of Dai-ichi Kangyo Bank in the 1970s, for example, was the second chairman of the now defunct Federation of Bank Employees' Unions (Zenginren) which was established just after the war.

The development of increased loyalty on the part of employees to the bank they work for has loosened the control of the industry-level organisation, the Japan Federation of Bank Employees' Unions (Zenginsoren), and labour-management relations have become more decentralised to the level of individual banks. Even at the level of the enterprise union, it has become more difficult to get an overall consensus of opinion among union members because, with the increase in the number of small branches, workers are so widely dispersed.

3.4.3 Effects on management

The power of personnel managers and middle-level managers has weakened considerably: their responsibilities have narrowed as branches have become smaller and more numerous. On the other hand, control over branches by central management, including setting targets, has strengthened. Management authority is becoming more and more centralised at the head office.

3.4.4 Patterns of negotiation and consultation

The banking industry has always offered white-collar employment, and, with its connection with capital and finance, labour-management relations have long been conservative and cooperative. This tendency has become even more pronounced recently.

Although the introduction of new technology has caused great changes in the banking industry, there have been no significant problems as regards resistance to or inability to use new technology on the part of unions or workers. At the end of the 1960s when on-line systems were introduced for the first time, there was some resistance from workers in labour-management consultations, but this quickly died down. Significant problems did not develop because workers realised that the services made available by the new systems were appreciated by customers, and that the workload was reduced.

4. Case-study of a machine-tool manufacturer

4.1 The context

4.1.1 The machine-tool industry

Machine tools used in metalworking can be loosely classified into two types. The first type is used to cut metal in order to produce the required specifications for finished or semi-finished products. These use electric motors for mechanical power. The second type is used to shape metallic materials by shearing, hammering, squeezing, etc., in order to produce products with the required specifications.

Metalworking is performed either with conventional machines or with the recently developed numerically controlled (NC) machines. Typical examples of the former include turret lathes, milling machines and jig-grinders. NC machines include computerised numerically controlled (CNC) lathes, machining centres, CNC jig-bores and FMS (flexible manufacturing systems). This study concentrates on NC or CNC machines, which have been widely introduced in recent years. Since NC and CNC are used almost as synonyms in the machine-tool industry, no distinction is drawn between the two.

The Japanese machine-tool industry used to lag far behind that of the United States and Europe in terms of technology. During the 1950s and the 1960s, however, a large amount of machinery was imported, and in the 1960s, production under licence from American firms began and domestic manufacturers started to develop their own technology. From about the middle of the 1970s, Japanese technology in the field of microprocessor-based CNCs has been recognised as the most advanced in the world.

Supported by the development of microelectronics technology, manufacturers of industrial machinery have expanded very rapidly. The performance of new machines such as CNCs has improved remarkably, and costs are lower than with conventional machines. Small and medium-sized firms used to suffer serious shortages of skilled labour and engineers; NC machines and industrial robots have therefore spread through various industries with unexpected speed – the car, electric applicances and metalworking industries, among others.

4.1.2 *The enterprise*

Company Z is a leading Japanese manufacturer of industrial machinery. Since the 1930s it has manufactured a wide range of machinery, including ships and boats, industrial machinery, pollution control systems and medical equipment. It has 6,000 employees; in addition to the head office, it has eight factories of various sizes and branches and offices all over the country.

Company Z used to specialise in heavy machines. In the 1970s, after the oil crises, demand for the firm's products was therefore sluggish due to the recession in heavy industry. Because of the circumstances mentioned above, however, the focus of manufacturing has gradually shifted to NC and other high-technology-based machines.

One characteristic of the firm is that it is both a manufacturer and a user of machine tools. In other words, it uses its own machine tools to manufacture machine tools. In addition, about 1,100 computer terminals have been introduced into the head office and establishments.

4.1.3 *Labour relations prior to the introduction of the new technology*

The present firm was formed from several large-scale mergers of machine-tool manufacturers, including two mergers after the war. Up to the 1960s, the labour-management relations of the constituent firms were integrated into the new company to a greater or lesser extent, making for a very complex situation. Moreover, there were many labour disputes since the firm's union belonged to an industry-based federation characterised by its policy of confrontation between labour and management. Eventually disagreement arose with this organisation over its expressed "opposition to improvement of productivity" simply on the basis of ideology. The union withdrew from the organisation in 1972, and the reorganised union adopted the usual enterprise union policy of cooperation with management. At present, the union belongs to the industry-level Shipbuilding and Heavy Machinery Workers' Union (Zosen-juki). Owing to the particular historical circumstances, however, a second union with fewer members coexists with the main union in some factories, each with its own distinctive ideology and policies. In spite of the generally cooperative relations between labour and management, the management are therefore very cautious, and pay close attention to both personnel management and labour-management relations. Labour-management negotiations in the firm are held at either the head-office or the establishment level, depending on the importance of the issue.

4.2 *The decision-making process*

4.2.1 *Decision-making regarding the introduction of the new technology*

Except for major new facilities that are very important for corporate strategy, most planning in Company Z is carried out by engineers at individual establishments in cooperation with head-office technical staff. In this way, difficulties that could arise, labour problems as well as technical problems, can to a certain extent be headed off at the planning stage, though they cannot be completely eliminated.

Depending on the scale of the new technology, its introduction is discussed and approved at the establishment, division or head-office level. The introduction of extensive new facilities or equipment is discussed by the board of managing directors at head office for final approval.

Since there is considerable choice about the types of technology or equipment that could be introduced, the firm's board of directors includes a number of engineers. This provides the specialist knowledge about new technology required for final decision-making.

4.2.2 The role of negotiation and consultation

Consultations between labour and management about the introduction of new technology are conducted at the head-office, establishment or workplace level, depending on the scale and importance of the technological innovation. Usually, proposals are presented and explained by management at the labour-management consultation meetings that are held monthly at the respective establishments and at head office. Plans are usually presented between six months and a year before the proposed time of introduction. At this stage, the effect the new technology will have on the workplace is explained, and detailed discussions are held between labour and management about the new facilities, including the time of introduction and the employment consequences. As the introduction of new technology often necessitates the redeployment and retraining of employees, the arrangements for this are the main subject for discussion. The management have not concluded a special agreement on new technology with the union, and up to now problems have been solved within the terms of the existing labour-management agreement.

A proposal to introduce new technology has never been completely rejected by the union, nor have decisions to introduce new technology ever been substantially modified because of union requests. In any case the management (the engineers in particular) are always well informed and hold full discussions about the technology. However, the union always submits requests regarding the redeployment of personnel caused by the introduction of new technology. These often concern the extent of redeployment, changes in working conditions, increased workload, other disadvantages caused by redeployment, and retraining (where it is held, how long it takes, and what programme is used). Details of the firm's original plan have often been modified in the face of such requests. Problems related to the introduction of technology or facilities have never led to serious disputes. If a serious problem arises, it can be fully discussed and solved through consultation.

4.3 Consequences for the workforce of introducing the new technology

4.3.1 Job security

The firm has never made employees redundant due to the introduction of new facilities. However, since one of the main reasons for introducing new facilities is usually improvements in productivity, it is natural to anticipate some labour-saving effect.

This was not a problem during the 1960s, when the Japanese economy was growing rapidly, and the firm was extremely successful; a great many temporary workers and workers dispatched from other firms were employed. During the depression in the latter half of the 1970s, on the other hand, the firm found it had too many workers, some of whom were on the regular payroll. The firm's main division, shipbuilding and related machinery, also suffered because of the depression in the shipbuilding industry as a whole and through the appreciation of the yen; drastic rationalisation, mainly through personnel cuts, was forced on the firm.

The firm tried to solve the problem by an immediate reduction in the number of temporary employees, but offering early retirement, and by loaning its employees to subsidiaries. The number of employees has decreased from 13,000, when business was booming in the first half of the 1970s, to the present 6,000.

As the pace of technological innovation accelerated, the firm adopted a corporate policy of reducing labour costs as much as possible while increasing investment in facilities. This does not mean, however, that there will be redundancies in the future: in order to maintain good labour-management relations, the firm intends to continue its basic policy of reducing the number of employees through attrition.

Furthermore, though the firm aimed at a decline in the workforce through attrition it could not halt regular recruitment for medium-term reasons. The firm had to develop new business areas due to the very rapid speed of technological innovation in general, and even during the period of depression it continued to recruit engineering graduates and male high-school leavers.

Redeployment of personnel was necessary in almost all cases where new technology was introduced, and was usually accompanied by retraining programmes for the workers. In this particular firm, union members cannot be redeployed to other establishments, so some establishments have had problems in the short term adjusting their workforce.

4.3.2 Work organisation

The labour union is worried that the gap is widening between older workers with conventional skills using conventional facilities and younger, less skilled (in terms of length of training needed) workers and engineers using the new facilities. The union requests at every labour-management meeting that the firm give particular consideration to this trend. At some workplaces, however, retraining workers with conventional skills has proved too difficult.

The content of work as a whole is changing as the new facilities diffuse, and "blue-collar workers" are becoming "white-collar workers". The motivation of workers to learn and participate in the firm's retraining programmes is becoming more marked. This is a reflection of the workers' increasing awareness that they themselves can expect benefits from improvements in the firm's business results.

The firm introduced total quality control (TQC) in 1967. At present, TQC activities account for two hours of each employee's working time every month. The management hope this will secure employees' commitment to their work, maintain high technical standards, and ensure good relationships between workers.

4.3.3 Payment systems and income protection

Generally, the introduction of new equipment causes various unexpected problems at the workplace, and this is true even when extensive consultations have been held beforehand. In the late 1970s Company Z changed its wage structure so that it was no longer based on seniority alone but also on job content and skill. This generated unusual situations at production workshops, such as workers who had completed high school being supervised by someone who had only completed junior high school, and sometimes there was smouldering discontent at the workplace for no good reason.

Moreover, the new wage system, according to which remuneration is closely connected to a worker's particular skills, tends to hold back the rate of increase in salary for middle-aged and older workers, who often experience difficulties in retraining, and take a longer time to master new equipment than younger workers, while younger workers can expect a rapid increase in salary. In this way, the meritocracy wage system can cause generational conflict within a labour union. The demand for higher wages is universal among the younger generation, who are gradually easing the reins of power from middle-aged and older workers.

Up to now, workers' income has not decreased after redeployment due to the introduction of new equipment because the union has always insisted on this. In the long term, moreover, benefits must develop for workers operating the new facilities, leading to differentials between them and workers still operating old equipment because the technological innovation was not introduced at their workplace. Since the union does not allow transfer of its members between establishments, blue-collar workers must to some extent depend on luck as to where they work. White-collar workers, on the other hand, are redeployed freely between establishments, so differences in work content cannot adversely affect their careers.

4.3.4 Training and retraining

The relevant retraining for the acquisition of new skills, which accompanies the introduction of new equipment, usually consists of short courses given at individual establishments by the firm's engineers or engineers sent by the manufacturer of the equipment. In other words, the workers who will operate the new system of production are selected, and on-the-job training is carried out by production engineers during installation and testing. In some cases, workers are sent to the equipment manufacturer for the training necessary to operate the new equipment, and they are sometimes expected to instruct other workers when they return. Younger workers who can quickly master new skills are usually selected for this task, which often makes older workers feel despondent. The workers selected to be at the core of a new production system must be in their late twenties to early thirties, with experience with general-purpose machine tools, a high skill level, the ability to adapt, and an interest in the new technology.

Recently, the operation of NC machine tools has come to require some basic knowledge of programming. Middle-aged and older workers who only have experience of manual machines have difficulties, and take a long time, adapting to the new systems.

4.4 Effects of the new technology on labour relations

4.4.1 Effects on the structure of the workforce

As mentioned earlier, the number of employees at Company Z has dropped drastically, almost halving since the early 1970s. Generally, workplaces at which new equipment has been introduced are characterised by a younger workforce, with the lower level of skill needed to operate the new equipment, and an increased number of engineers. The number of blue-collar workers in the production division has decreased drastically, while the number of white-collar workers has increased.

4.4.2 Effects on the unions

As in other Japanese firms, the union has a union shop, so when the firm is expanding and the total number of employees is increasing, the number of union members increases almost automatically. When the number of employees decreases, on the other hand, the number of union members decreases almost proportionally. As a result, in firms like Company Z which are being rationalised, the union shrinks and is often thrown on the defensive. In Company Z, moreover, the decrease in the number of blue-collar workers in the production division and the increase in the number of white-collar workers has caused changes in union policy, because in general white-collar workers are less involved with the union.

4.4.3 Patterns of negotiation and consultation

Until the latter half of the 1960s, the firm failed to establish cooperative labour-management relations, judging from the frequency of strikes and labour disputes. However, as stated earlier, the situation changed, and although there do remain some possible flashpoints, labour-management relations have become fairly stable.

One bone of contention is the continuing trend towards meritocracy-oriented management. This is alienating middle-aged and older workers, who are expressing doubts about the fairness of the system of evaluating workers' achievements. Many workers feel they would have less anxieties under a wage system based on seniority. This meritocracy-oriented management policy, along with the technological innovations, may to some extent have destabilised labour-management relations. At present, the union does not seem to be able to deal effectively with such complaints at the workplace level, which is another factor leading to decreased participation in union activities by the ordinary members.

Generally, manufacturers of machinery such as industrial robots and NC machine tools are having good sales results in spite of the recent general recession, due to the high value of the yen. The firm is now diversifying from being a manufacturer of shipbuilding machinery, but overall business is still not very good. Recently, labour-management relations have been calm because the rationalisation is now complete. The future of labour-management relations depends on the firm's business operations from now on, but it is difficult to be optimistic.

5. Evaluation

Although the printing, banking and machine-tool industries each have peculiar characteristics, all three have clearly been affected by technological innovations based on microelectronics.

Business conditions in the printing industry as a whole are favourable because of the expansion of demand with the coming of the information age, although there is severe market competition. The firm selected for the case-study has a large market share, and is known for its high level of technology. Both labour and management have a relatively flexible attitude to new technology, making possible the unique system in which old and new facilities exist side by side. Although the workforce is generally fairly old, it can introduce new technology without serious problems. In contrast to large firms, it is thought that small and medium-sized printing firms (of which there is a large proportion in terms of numbers) are having more difficulties because they lack the ability to develop technology themselves. Moreover, small firms are not able to adjust labour by redeployment, and their operations still rely on workers with the traditional skills.

The average size of banks is very large, and the bank studied here is no exception. Due to deregulation and intensified competition, rationalisation based on technological innovations such as office automation equipment has developed on a large scale and at a very high speed. Among Japanese industries, banks are thought to be enjoying good business results, and can provide good, stable employment opportunities. The bank studied here has had very stable business operations. However, conditions in banks are expected to become less favourable in many respects. Rationalisation, accompanied by the introduction of new technology, has recently been changing the workplace dramatically. With the further introduction of automation, the working environment will become very different from the conventional labour-intensive conditions. However, except for a short period after the war, labour-management relations in banks have been very cooperative, and, as far as the foreseeable future is concerned, it is considered that labour-management relations in this industry will not change much.

In the machine-took industry, firms specialising in, for example, industrial robots have had very favourable business results. However, firms like the one studied here, which originally stressed other areas of business and are now trying to shift to the field of machiner tools, cannot be optimistic about growth. The firm needs to scale down its original area of business through rationalisation, while shifting major resources (capital and labour) to more promising areas. If the firm introduces the system of meritocracy too quickly in order to adopt new technologies, there may be friction between younger workers and older workers possessing the traditional skills. Moreover, the firm has considerable experience of confrontational labour-management relations, and it must take a prudent attitude towards rapid change. In this type of firm, even the term "cooperative" has quite a different meaning to the one it has in the printing industry.

The common features of all three case-studies are that competition is intensifying, and that both labour and management are more interested than ever in the growth and stability of their firms. In order to strengthen the competitive power of their firms, workers are therefore in principle very cooperative about the introduction of new technologies, provided these technologies do not cause a deterioration in working conditions.

In addition, for some time, the centre of gravity of labour-management relations (that is, the place for decision-making) has been shifting from the industrial level to the corporate level, and further to the establishment level, or even the workplace level. This general trend may be termed the "fragmentation" of labour-management relations.

Notes

[1] Apart from documents in Japanese, see *Proceedings of the International Symposium on Microelectronics and Labour* (held in Tokyo during 25-27 September 1985) as a useful publication in English giving a relatively intelligible overview of the issue.

[2] One good case-study is Mikio Sumiya (ed.): *Gijitsu-Kakushin to Rôshi-Kankei* (Tokyo, Nihon Rodo Kyokai, 1984; in Japanese). It contains an analysis of 15 industries, including the car, newspaper, printing, software, electric and steel industries.

[3] See, e.g., Y. Kuwahara and F. Kimura: "Gijitsu-Henka to Koyo", in *Journal of the Japan Institute of Labour*, Feb. 1983, pp. 38-48.

[4] Y. Kuwahara: "The industrial locus of trade unionism", in *Japan Labour Bulletin* (1983); Y. Kuwahara: *The job creation and job destruction process in Japan*, submitted to the OECD's Programme on Job Creation and Human Capital Investment in the Context of Technological and Other Structural Changes (Tokyo, 1986).

[5] According to the annual "Survey of Establishments" published by the Statistics Bureau (Somucho), an extremely large number of small firms exist in printing and related industries. If "small" is defined as having 30 employees or less, 95 per cent of firms in the printing industry are small, as opposed to 92 per cent in manufacturing industry as a whole. This characteristic of the printing industry is clearly demonstrated by the fact that the percentage of those employed in small firms is 38 per cent in manufacturing industry as a whole and 61 per cent in the printing industry.

5 Sweden

Technological change and labour relations in Sweden

Bernd Hofmaier in collaboration with Kristina Hakansson**

1. Labour relations in Sweden

1.1 Labour laws and agreements

As early as 1906 the first labour-management agreement was signed by the Swedish Trade Union Confederation (*Landsorganisationen i Sverige* – LO), the largest union confederation, and the Swedish Employers' Confederation (*Svenska Arbetsgivareföreningen* – SAF), the largest employers' organisation. The next few decades saw various labour laws and labour agreements, and a continuation of labour-management cooperation. By the 1950s and early 1960s a "Swedish model" had taken shape, characterised by centralised collective bargaining and an active government labour market policy. In this period there were relatively few labour disputes, but during the second half of the 1960s the labour market became increasingly unruly.

Within a few years, major portions of existing legislation were revised and a wide range of new labour legislation adopted. This can be seen as a response to increasingly pressing demands for a better work environment, improved job security, more democracy in working life, improved working conditions for the trade union organisations, and so on. A so-called Labour Law Committee was established whose work culminated in the Co-determination Act (MBL) and the Public Employment Act (LOA), both introduced with effect from 1977. According to one member of the committee:

> The Co-determination Act ... is based on the established negotiating traditions of the labour market, which had already resulted in employees and their organisations gaining influence on other matters beside pay and similar material terms of employment. The Act does not aim at any fundamental transformation of the balance of power in the labour market. The declared intention of the legislature was for the labour market parties themselves ... to progress further, extending co-determination by means of clauses in collective agreements.[1]

The drawing up of a Development Agreement between the Swedish employers' Confederation and the employee unions was assumed in the co-determination legislation but came about only after considerable conflict and delay. The

* Swedish Centre for Working Life.

Agreement on Efficiency and Participation (UVA) was signed between the Swedish Employers' Confederation, the Swedish Trade Union Confederation and the Swedish Federation of Salaried Employees in Industry and Services (PTK) in 1982.

The agreement starts out from the general premiss of mutual understanding of the need for efficiency, profitability and competitiveness in enterprise, this being a requirement for employment opportunities, job security and development at work. It includes clauses concerning the goals and direction of joint development work, forms of co-determination, the role of local agreement in small firms, the giving of information to members of the local union during working hours, employee consultants, and rules for negotiating procedures. The agreement is similar to the Co-determination Act in that it provides a framework which presumes that agreements and implementation rules will be added in order further to specify the legal requirements, at both industry and enterprise level.

Two other important pieces of legislation are the Act on Security of Employment of 1974, which stipulates that employees are normally employed on a permanent basis, and the Work Environment Act of 1978, which increased the rights of trade unions to help improve the working environment. The concept "working environment" encompasses new work systems, working hours, and the adaptation of work to human factors, both physical and psychosocial.

1.2 Swedish unions and technological change

Traditionally most Swedish unions have been and still are positive in their attitudes to technology and technological rationalisation, even in the 1930s when one-third of union members were unemployed – possibly due to a belief that rationalisation and new technology and the resultant higher productivity would lead to a better economy and create new jobs.

Labour and management are in broad agreement on their approach to the development and utilisation of new technology. In the spirit of the Agreement on Efficiency and Participation, it is recognised by both parties that the introduction of new technology should take into account the differing interests of all the parties involved. New technology should be utilised to maintain and further develop the knowledge and skills of individual workers, to facilitate new forms of work organisation with new scope for worker participation, and to eliminate environmental risks. Workers and their unions should play an active and meaningful part in the development process.

These goals were set out in *A new world of work* (1988), produced by the Development Programme for New Technology, Working Life and Management, which was launched in 1982 by the labour market parties in collaboration with the Swedish Working Environment Fund. Its aims were to investigate the possibilities of improving job content and work organisation, as well as promoting high productivity and competitive capacity, in conjunction with the introduction of new technology. Roughly two-thirds of the programme budget went directly to 25 projects in the manufacturing and service sectors.

In 1985 the "Swedish Working Environment Fund's Programme for Research and Development in Leadership, Organisation and Participation" was set up. It was to concentrate upon organisational development, and aimed to utilise research in support of practical development in working life.

A third programme, launched in 1987, is the so-called MDA programme, where MDA is the Swedish abbreviation for Man, Computers and Work. This programme is more computer software-oriented, and is financing, among other projects, research on user-friendly computer systems.

The basic task is thus to develop new technology that serves both the objectives of industrial policy and the objectives that are central to a policy on working life. These objectives are in competition with each other, but this can be overcome through government and union involvement in the development of technology.

Another key assumption is that the trade union organisations are capable of expressing the demands of employees. They should also be capable of drawing upon the employees' know-how and experience of work, resulting in a joint development of technology that satisfies the demands of the users. The means for achieving this are, among other things, the Agreement on Efficiency and Participation or the so-called Development Agreement of 1982. Given strong unions and a traditionally calm labour market, this would give Swedish industry a competitive advantage in an increasingly turbulent world market.

2. Case-study of a national newspaper

2.1 The context

2.1.1 The newspaper industry

There are about 150 newspapers in Sweden. About 90 of them publish at least four days a week. As in many other West European countries, the number of newspapers has greatly reduced recently, especially during the 1950s. In spite of this, total sales have increased, though the rise is concentrated in a few newspapers. Today, 15 newspapers account for half of total sales. The average sales figure for all newspapers is about 45,000 copies a day.

By international standards, Swedish newspapers enjoy very extensive subsidies, both direct and indirect. The most predominant form of direct assistance is a so-called production subsidy which goes to newspapers that have no more than 50 per cent of the market share at the place of publication. Until the 1950s, it was held that newspapers should compete with each other freely without intervention from the State. After the closing of two major newspapers in Stockholm and Gothenburg, the classical liberal ideology changed slowly. A government commission reported in 1968 that several other newspapers were in difficulties. In 1971, the Social Democratic Government recommended the use of selectively directed subsidies for the production of newspapers. In 1976 a development subsidy was introduced for newspapers that wished to invest in new technology and wanted to reorganise their printing offices.

During the 1970s, newspaper technology changed drastically. The most striking change came in the composing room, where computerised photosetting replaced hot-metal setting. This means that much of the typographer's knowledge has been programmed in computers and the character of handicraft is lost. Electronic systems also play an important role in modern offset printing techniques, but they did not create the same revolution as that which affected the composing room.

2.1.2 Labour relations in the newspaper industry

The Graphic Workers' Union (*Grafiska Fackförbundet* – GF) was founded in 1973 through the merging of three different graphic unions, the typographers', the bookbinders' and the lithographers' unions. Membership has increased in the last few years and is now about 60,000. The union includes three large professional groups, compositors, printers and bookbinders, and one small group, reproduction workers.

The Graphic Workers' Union signs collective agreements with five different employers' organisations, including the Graphic Employers' Association (GA) and the Newspaper Employers' Association (TA). These regulate relations between the parties and include clauses on settlement obligations, application of agreements, method of conciliation, etc. They also indicate the method of deliberation between the concerned parties. The employers' association has an obligation to the union to consider carefully all the implications before decisions are taken about important technical changes. There is also a collective agreement about wages. It regulates among other things the placing of workers in salary grades, minimum wages, and compensation for inconvenient working hours.

The Graphic Workers' Union also has a technical agreement, signed by the Graphic Employers' Association and the Newspaper Employers' Association, and a cooperation agreement concerning technology with the White-collar Workers' Union (GFT) and the Swedish Foremen's and Supervisors' Association (SALF). With the Swedish Journalists' Union there is agreement on a joint line of action and on the division of work when newspapers introduce new technology.

The technical agreement with the employers' association was signed in 1974. It stipulates that employment security and salaries are not to deteriorate when new technology is introduced. Nobody can therefore be unemployed because of new technology. In fact, this provides greater employment security than that provided by law. The agreement also stipulates that, in cases of important changes in the workplace, the union has the right to participate in project groups.

The cooperation agreement with other unions is to make clear the demarcation between the different professions when new technology is introduced. In concrete terms it means that printing is a graphic workers' profession, whatever technology is used.

The implementation of the new technology led in many cases to conflicts, especially at the two national newspapers, *Dagens Nyheter* and *Svenska Dagbladet* in Stockholm. At the *Dagens Nyheter* there has been conflict not only between management and union but also between different unions, mainly jurisdictional disputes as to who will carry out the new work tasks after implementation of the new technology. At the *Dagens Nyheter* these disputes have led to strict labour contract regulations. At the *Svenska Dagbladet* jurisdictional disputes have been dealt with in a more informal manner in accordance with the intentions of the Development Agreement.[2]

The smaller newspapers have mostly dealt cooperatively with the introduction of new technology, and the jurisdictional disputes that have occurred, for example between printers and journalists, have usually been handled in a constructive way.

2.1.3 The enterprise

Göteborgs Posten, located in Gothenburg, is today the second largest morning paper in Sweden. It sells about 300,000 copies on weekdays and 350,000 copies on Sundays. The firm also prints *Göteborgs Tidningen*, which is one of four evening papers in Sweden.

Göteborgs Posten and the firm that publishes the two papers is owned by the Hjörne family. In contrast to the situation in the majority of the privately owned newspapers, the family retains the position of editor-in-chief, thus combining ownership with direct influence on the content of the paper. In the early 1980s the firm changed in structure, and it now consists of three divisions: *Göteborgs Posten*, *Göteborgs Tidningen* and a technical division.

The technical division has about 430 employees and is divided into six sections:

(1) The composing room, with approximately 200 compositors. Here the text is written in and the pages assembled; advertisements are created and put together.
(2) The reproduction department, with 35 employees, responsible for touch-up, creating pictures and colour assembly.
(3) Process engraving, with 25 employees. Here the offset plates for printing are made.
(4) The pressroom. It is here that the newspaper is printed, using the new offset technology. The main work tasks are to supply the printing-press with paper, assemble the offset plates, regulate colour input, and supervise the printing process. After reorganisation there were approximately 90 employees.
(5) Distribution, with 80 employees, responsible for bundling newspapers and taking them to vans.
(6) The service section, with six employees who are responsible for maintenance and repair for the entire technical division.

The occupational groups in these sections are compositors, photographic operators, printers, reproduction workers, distributors, service personnel and maintenance personnel.

The object of this case-study is the pressroom. Having used the relief technique for printing for many years, *Göteborgs Posten* has now introduced the offset technique, which makes colour printing possible. With the relief techniques, only black and white could be printed.

Computerised equipment in the pressroom is used for three tasks:

(1) supervising the stretching of paper to reduce the number of paper breaks and thus increase the speed of total production;
(2) obtaining a more uniform and improved quality;
(3) making different kinds of prior adjustments.

This new technology did not change the work in the pressroom alone: it produced a sort of chain reaction so that work in the composing room, the reproduction department and process engraving all changed. Even distribution received new machinery to keep pace with the effects of the new printing technology.

The firm bought four offset printing-presses. The installation took place in different stages and today three of them are in operation. The fourth is planned to start in April 1988. In all, it will have taken two years to install all four presses. During this time both relief printing and offset printing will have worked in parallel.

2.1.4 *Labour relations prior to the introduction of the new technology*

At the local level the enterprise union (*Grafiska Fackklubben*) at *Göteborgs Posten* includes all 430 workers in the technical division. The local union board has nine members representing all occupational groups in the six sections. In addition, the three largest sections, the composing room, the pressroom and the distribution section, each have their own small union board. Their function is to be a working team for each respective section. They handle the problems of their section and propose measures. The formal decisions are then made in the local union board. The chairman of the board works full time on union tasks.

Even the foremen are connected with the local Graphic Workers' Union, which is quite unusual compared with other newspapers. It is thus the only trade union in the technical division.

2.1.5 *Reasons for introducing the new technology*

There were two reasons for buying new printing-presses. First, a product of better quality and colour printing were both necessary. *Göteborgs Posten* is to a great extent dependent on advertisements, and the advertisers were therefore indirectly pushing for colour printing. Second, the existing machinery was old and worn out.

When the decision to buy new presses was taken, both directors and union agreed that the offset technique was the right one to choose.

2.2 *The process*

2.2.1 *The initiative*

The firm moved to a new site in the late 1970s. At that time, the employees suggested that the firm buy new printing-presses. The building was, however, planned for the old machinery and the directors turned down the suggestion, for financial reasons.

In the early 1980s the firm made no profit, and the directors decided that new technology was necessary to obtain a better quality product and to make colour printing possible. In 1981 the board of directors decided to buy new printing-presses. The union was then informed according to co-determination law and the joint project groups started work.

The first initiative had in fact come from the employees. New presses were discussed for a couple of years before the directors made the necessary financial decision.

2.2.2 The role of the project groups

Three sections – the process-engraving department, the distribution section, and of course the pressroom – were directly affected by the installation of new printing-presses. One project group was therefore set up for each section. In reality, the pressroom dominates the other sections and its project group was therefore the most important. This group included four people, two representing management and two the union. Apart from the project groups, there was also a consultative group, in all 18 people. The entire union board from the pressroom and the industrial safety representative participated in the consultative group.

The project group's task was to suggest which press the firm should buy. The suggestion had to be accepted by the consultative group, after which it was put to the firm's financial unit. The purpose of the project group was consequently to ascertain which technical system would fit; the project group was not at all involved in the economic decision.

There are about five manufacturers in Europe of this kind of offset machinery. The project group asked for tenders and for representatives of the manufacturers to come to Gothenburg and describe their systems. The group then visited a couple of newspapers in Europe to see the presses actually working. The consultative group also participated in some study tours. After two years, the project group delivered a final recommendation to the firm's management. The consultative group and all personnel in the pressroom had approved this suggestion so it was well received by the employees. The essence of the final conclusion was that three printing systems were equally good from a technical point of view. It was then up to the finance unit to take up the issue with the manufacturers.

The project group presented a list that included both technical and work environment demands. First and foremost was the demand that the manufacturer of the printing-press should be an established firm. It must be possible to obtain service and spare parts in the future. The union also made a request to buy Swedish. The request had to be taken back when it was revealed that Swedish manufacturers did not have the capability to build a printing-press of the size required.

The most important technical requirement was efficiency, that is, the shortest possible production time and dependability. The work environment demands included a low noise level. The employees' workspace lies in close proximity to the printing-press. The noise level there should not be higher than 65 decibels, that is, speech level. Other demands related to safety equipment, easy access to emergency stop devices, and so forth. The project group also agreed on certain ergonomic demands. For example, the group requested that all the adjustment mechanisms on the printing-press should be readily accessible. As regards this demand, however, consideration would also have to be given to the specific workspace area.

As regards training, the union asked that about 25 printers should go abroad for training as there was no equivalent printing-press in Sweden. This demand was regarded by the management as impractical as normal production would have to be maintained. After further negotiations a compromise was reached that 11 printers would go abroad for training.

The printing manager and the head of the local union were both in the project group and regarded themselves sufficiently qualified to make most of the technical decisions without resorting to outside consultation. This meant that the group did not have to carry out further research. The project group saw themselves as well able to evaluate the technical specifications that the suppliers presented.

2.3 Consequences for the workforce of introducing the new technology

2.3.1 Job security

The introduction of the new printing technology definitely did not lead to any reduction in the number of employees. According to the union, it has led to a need for more personnel. During the period of introduction and training of the ordinary personnel, deputies were engaged. This seems to be a permanent need. The new technology makes it possible to produce products of better quality. The improvement in quality leads also to demands for more crew members. Consultations about additional employees are therefore being undertaken.

There are also ongoing consultations concerning the minimum number in a work crew. The union has demanded seven workers for every printing-press. In cases of long-term absence a temporary replacement should be employed. If an agreement about minimum numbers in crews can be signed, it will create a precedent for other newspapers: the local unions are therefore in contact with the central union as regards this question.

2.3.2 Work organisation and working conditions

The work tasks have changed. In the pressroom both preparing the product and supervising is conducted by computers. There are three computer terminals to every press. With their help the printers can adjust colour, damp and contrast. More extensive preparation work and more cleaning work after printing is completed are two of the results. The new presses are built in and the workers can change and adjust the printing process by computers from a central cabin. The relief printer has to be adjusted manually. The refilling of colour used to be done manually by buckets, but now it works automatically.

Workers in the pressroom belong to three different work categories with different degrees of responsibility. The three categories are "first-class printers", "second-class printers" and apprentices. The first-class printer has the major responsibility for the printing, that is, for ensuring that the newspaper is produced in the right quantity and quality. The second-class printer helps look after the adjustment and regulation of the printing-press. The apprentices feed paper into the paper rolls and make sure the paper flow is maintained. On their own initiative, the workers rotate between these positions to avoid the creation of a static hierarchy. This form of organisation and rotation has not changed with the introduction of new presses.

The differences between these three work categories will probably disappear in the long run. The goal is that everyone who works in the pressroom should be a printer. There is, however, a shortage of offset printers and the firm must therefore train newly employed personnel.

Mobility, both inside the firm and between different firms, has not changed since the introduction of offset printing presses. Nor has the possibility of promotion changed.

The work environment (noise, dust, dirt, etc.) is much better since the introduction of the new presses.

2.3.3 Payment systems

Salaries are paid monthly, but are based on a week's pay plus a bonus for inconvenient working hours. The gross salary is, however, equalised over the whole year so that the pay is the same every month. As a rule an individual works as an apprentice for the first year and is paid basic wages. After a year one moves into the "larger wage group". A "first-class printer" earns a bit more. The difference is not great. There are no other bonuses. This pay system did not change because of the new presses.

2.3.4 Training and retraining

One of the project group's tasks was to look at training. When the choice of press was made, all workers had four weeks' training. The press used was much smaller than the one the firm bought, but it provided knowledge about the new printing principles. The new technique also demands much more care. The employees did not think the training they received was sufficient, partly because the training took place over a year before the press was installed, and partly because the new press was much more complicated than the one used for the training.

Both management and union agreed that it was very hard to plan the training before the new presses were installed. There was no similar press at any other newspaper in the country. However, as the new presses were installed in stages, it was possible to work with both the old and new equipment in parallel. In this way it was feasible to train the personnel and change over gradually.

About ten of the workers also went abroad for training with a similar press. These workers then trained the others in the pressroom. Two instructors were provided by the manufacturers for training the workers. Some further training is intended, and there is also consultation going on about training in colour printing.

When the printing department received the new printing-presses many other sections were affected. This was something the project group had not anticipated, despite the fact that the plate assembly and distribution sections received some new equipment with the arrival of the new printing-press. As a result there were never any plans for the training of personnel in the other sections.

2.4 Effects of the new technology on labour relations

At the time of the introduction of the new technology, labour relations at *Göteborgs Posten* were good. There had not been any significant conflicts, nor were any caused by the introduction of the new printing-presses. As the chairman of the union says, working at a newspaper means deadlines every day, and everyone does his best to deliver the paper. This is largely a matter of teamwork, and the workers do not think much about relations between management and employees. This is typical of the newspaper industry.

The pattern of consultations followed the intention of the Co-determination Act. As described earlier, this Act requires the employer to take the initiative in negotiating with the local employees' organisation before deciding to effect any major change in working conditions or activities. According to the local union at *Göteborgs Posten*, the union representatives received sufficient information in the beginning and under the project.

Local co-determination may take various forms, according to the Co-determination Act. At *Göteborgs Posten* the union took part in the process of introducing new printing-presses by participating in the project group and in consultations. The intentions of the Co-determination Act (MBL) and the resulting Efficiency and Participation Agreement (UVA) were thus followed.

3. Case-study of a bank

3.1 The context

3.1.1 The banking industry

The Bank of Sweden (Sveriges Riksbank) is the national bank that issues banknotes and determines discount rates. There are 15 commercial banks with about 1,500 branch offices located throughout Sweden. There are also 149 savings banks with about 1,400 branch offices, and a cooperative bank which consists of 392 credit societies acting through some 700 branch offices. About ten foreign banks have offices in Sweden.

Swedish banks are among the most advanced in the world with regard to technological equipment. According to *The Economist*, "The Scandinavian banks have plunged into the latest in automation with wild enthusiasm ... Scandinavian banks now have the highest level of spending on computers per employee of any country in Europe" (12-20 March 1981). This was written at the beginning of the 1980s but still reflects the situation today.

Computerisation in Swedish banks has been carried out in several steps. The first step was the transition from keeping customers' accounts on paper to electronic storing. The next was transfer between banks, and the third is the introduction of a new system for electronic payment transfer. Computerisation has led to:

(1) more routine tasks being done by the customers themselves, for example using automatic machines for withdrawals;

(2) decentralisation of lending operations, foreign business, operations with bonds and shares, analysis of companies, and marketing;
(3) extension of services offered by banks, for example financial guidance, tax assistance and tax planning.

In the 1970s computerisation actually led to an increase in the number of employees in Swedish banks. The opening up of new markets for services is one explanation, especially since banks began to take over firms' wage transfer operations. Another expanding sphere of activity is foreign business. Swedish firms have extended their international business more and more, and Swedish banks have constantly adjusted and developed their service to the firms' needs.

In concert with computerisation there have been changes in the work. In general it can be said that there has been a polarisation of work tasks, with some employees having more demanding work tasks, while others have more simple and routine ones. It is mainly women that do the low-status cashier work. Present tendencies indicate that for the most part cashier work will disappear. As a consequence, training is being provided to enable the personnel in question to take on other tasks.

In general different banks have had similar motives for introducing new technology. Competition among banks had become more severe, and the competitive strategy was to improve customer service. Personnel directors realised that they were losing qualified people because of obsolete computer systems. One means of improving customer service and attracting people was to introduce new and more powerful systems.

3.1.2 Labour relations in the banking industry

There is a co-determination agreement between the Association of Swedish Banking Institutions (BAO) and the Swedish Bank Employees' Union (*Svenska Bankmannaforbündet* – SBmF). The aim is to facilitate the solution of problems harmoniously and without undue waste of time. It stipulates that co-determination should be incorporated into banks' regular decision-making structure, and that employees have the right to representation on all decision-making bodies. Information shall always be provided before decisions are taken with respect to important changes in the conditions of work.

There also exist joint recommendations concerning technological developments on the part of the BAO and the SBmF relating to staffing, training, personnel development and similar issues. The parties agree that work shall be organised in such a way that VDU work is only part of an employee's job. There is also an agreement about regular eye examination and health checks.

Savings banks have their own central organisation for dealing with technological change, with trade union representatives from different savings banks throughout Sweden, but the commercial banks have no similar organisation. Instead, an individual bank will set up a steering group, or project group, with representatives from both management and union, to discuss issues relating to technological change. This group does not, however, make decisions, which are made in a so-called administrative committee, again composed of representatives from both management and union. Issues that the steering group cannot agree upon are directed to this administrative committee.

Usually all branches of a bank are affected by the introduction of new technology. As far as the introduction of computers is concerned, each branch can determine its own needs. The technology that is appearing now can be used at both local and central levels, which means new demands for union participation at both these levels.

3.1.3 The enterprise

Bohusbanken is the smallest commercial bank in Sweden, with one office in Gothenburg and one in Stockholm. The bank has customers all over Sweden and concentrates on establishing itself as the main bank for small and medium-sized firms. This bank is part of the Wallenberg finance group, which controls a great part of Sweden's most important manufacturing industry, and therefore has contact with the really large firms by being their second, third or even fourth bank. The bank has an investment branch in Stockholm and staff at the stock exchange there. The head office is located in Gothenburg. All statements of accounts, bookkeeping and administration are done there.

Bohusbanken has expanded significantly during the last few years, from 15 employees at the head office in 1979 to 60 today. The employees are all economists, divided into two professional groups, those with university training and those with upper secondary school education.

The bank has a board of directors, a management unit for questions about policy, and a credit committee which handles all new credits. It is divided into four sections:

(1) the foreign business unit;
(2) the capital funds unit;
(3) the credit unit; and
(4) the lending unit (for minor customers).

The bank computerised the foreign business unit in 1984 as a member of the Society for Worldwide Interbank Financial Telecommunication (SWIFT). SWIFT was established in 1979 by the major banks in the world as a company owned by its member banks. Each bank has a share in relation to its size, and pays for its transfers. The purpose of SWIFT is to make fast communications possible, using electronic equipment instead of telex. All transfers of payments and information are done electronically. The majority of banks in the Western world are members.

3.1.4 Labour relations prior to the introduction of the new technology

About 90 per cent of the staff from both professional groups are connected to the Swedish Bank Employees' Union (SBmF). The local union board consists of four people, none of whom works full time on union tasks. As the board is quite small, it is possible to be flexible and hold meetings at short notice. The most important task for the local union is wage negotiations.

3.1.5 Reasons for introducing the new technology

There were three reasons for becoming a member of SWIFT. First, marketing: as SWIFT is used by the major banks in the world, being a member has a certain status attached to it, and that is important in the case of new commissions, especially for a small bank. Second, effectiveness: SWIFT is extremely effective as regards payments and information. The system also permits more accurate and safe handling than manual handling. Third, expansion: the system was a precondition for further expansion, an "investment in the future".

3.2 The process

3.2.1 The initiative

The discussion about membership in SWIFT started in 1979 when the foreign business unit was set up. The management thought, however, that it would be too expensive and postponed the decision. Later, the manager of the foreign business unit and his assistant, who was also a shop steward, suggested that the bank should be a member of SWIFT. The shop steward was a very influential force for this development and the foreign business unit was positive.

In 1983 the board of directors decided to investigate membership in SWIFT. The shop steward was also a staff representative on the board and therefore got information about the computerisation before any decision was taken. A project group was then appointed.

3.2.2 The role of the project group

The project group consisted of three people, two management representatives and the shop steward who had the main responsibility for the project. The shop steward was partly released from his ordinary job for work in the project group. The aim of the project group was to calculate the costs of the computerisation, both financially and in terms of employment. The project group worked completely independently of the central management. The three members of the group were in complete agreement about their goals and worked together to reach them. There was no choice between different computer systems. The choice was whether or not to become a member of SWIFT. The group's conclusions were readily accepted by both management and union; formal negotiations were only held about training.

The project group worked entirely at a local level. Bohusbanken is a very small workplace where the manager of the foreign business unit and the shop steward work closely together. Both management and union are positive towards computerisation, and the union does not regard it as any threat to employees. Management and union also agree that membership in SWIFT corresponds well with current needs. The system promoted expansion, and there was simply no alternative to membership. That explains how it was possible to introduce the SWIFT system in such a short time: there were no obstacles.

3.3 *Consequences for the workforce of introducing the new technology*

3.3.1 *Job security*

After computerisation, the bank expanded further and new people were employed, so no jobs were lost as a result of joining SWIFT.

3.3.2 *Work organisation*

About ten people are directly affected by SWIFT. The work tasks are principally the same, but the techniques of bookkeeping and making payments are changed. The work tasks are now concentrated on terminals. The staff do not experience this change as entirely positive. As a result of negotiations, they have regular eye examinations and no pregnant women are obliged to work with VDUs.

SWIFT is a closed system, so the employees have no influence at all on the internal programme structure. On the other hand, employees received training in an area about which they had no knowledge before computerisation.

Computerisation has not led to any changes in promotion prospects. People who work with SWIFT are now more competitive, however, and that could lead to greater mobility between banks.

3.3.3 *Payment systems*

Employees receive different basic salaries depending on their level of education. The basic salaries are guidelines, settled by central negotiations. The employee's place in a salary group is then decided by negotiations at a local level. Accordingly there are negotiations both about individual salaries and about salaries for groups of employees. There are also local consultations concerning a given share of the total salary figure in the bank, put aside for employees who are particularly diligent. The salary system has not changed since computerisation.

3.3.4 *Training and retraining*

The shop steward on the project group took the initiative in demanding that all personnel in the foreign business unit should take a general computer course. This went on for three evenings a week after work hours for a six-month period before SWIFT was introduced. Employees also received training from a specialist in the SWIFT system from Brussels. This included a general course in SWIFT and two specialised courses, exchange dealings and document handling. Each course lasted a couple of days.

3.4 *Effects of the new technology on labour relations*

Computerisation has not led to any real changes in union or management. Relations between the parties are characterised by a strong tradition of cooperation and unity. No conflicts have arisen. Management and employees work closely together and contact is of an informal character. This situation has not changed since the introduction of SWIFT, and computerisation was introduced in complete harmony.

The bank has not established any consultative group. After computerisation, however, the bank did start weekly information meetings. All staff participate in these and the purpose is both to inform and to avoid conflicts. Without doubt there is a more informal relationship between management, union and employees at the Bohusbanken than there would be in a larger bank. Many questions are dealt with informally and decisions evolve from discussions. In practice co-determination does exist even though formal negotiations do not take place on all issues.

4. Case-study of a machinery manufacturer

4.1 The context

4.1.1 The enterprise

Svenska Kullager Fabriken, or the SKF group, is one of the largest industrial groups in Sweden. SKF is the world's leading manufacturer of ball and roller bearings, with about 20 per cent of the total market. The principal purchasers are the car industry, domestic appliance manufacturers and makers of home electronic equipment. Over 80 per cent of the total business is in bearings, about 11 per cent in component systems, and about 6 per cent in tools. Net sales in 1986 were 19,967 million kronor and the whole group has about 44,800 employees, of whom about 4,000 are in Sweden.

SKF did quite well during and after the Second World War. In the 1960s problems arose concerning a shortage of workers, and SKF recruited workers from Italy, Yugoslavia and Finland. The 1970s was a dramatic period. For the first time in SKF's history, the existence of the firm was threatened. There were several reasons for this: increased competition on the bearing market, especially from Japanese manufacturers; old production equipment and low profits; and an extremely bad situation on the world market. The later part of the 1970s was the most depressed since the 1930s. During this period the workforce declined by about 18,000, 1,800 in Gothenburg alone.

In the 1980s the firm went through a major structural reorganisation of the production programme. The aim was to coordinate the production programme of the different establishments in Europe, the intention being that each production unit should only produce one bearing type, and that an automated production and control system would lead to reduced production costs.

This coordination between the different establishments demanded a powerful planning system. SKF therefore introduced a new so-called "Global Forecasting and Supply System" (GFSS), which is still in use. For the establishments in Gothenburg this system resulted in a change of products and in a new production concept with line production, shorter lead times (time between initiation and completion of the production process) and greater flexibility. Production was also to be more tailor-made to customers' requirements.

4.1.2 The development project at D3

A new production concept with these characteristics was therefore planned in part of the Gothenburg establishment, the so-called D3 plant. In the D3 plant, bearings with a diameter of 180-240 mm were manufactured with traditional machines. About 200 employees worked here grinding, polishing and assembling roller bearings. The old machines were responsible for quality problems, high production costs and a poor work environment. Numerous minor rationalisation attempts had been made, and in 1980 these were put together under one roof with the project title "Concept 1-4". This concept envisaged a totally new production organisation with a flexible manufacturing system (FMS) consisting of several minor flexible manufacturing cells. The operators' job would be to supervise the process, but their involvement would be limited and they would need only limited qualifications. There would be ten parallel straight lines in which the working and assembly of ball bearings would take place. The accomplishment of "Concept 1-4" would clearly take place over a long period of time.

From the very beginning there was opposition to the project within the local branch of the Swedish Metal Workers' Union and of the Swedish Union of Clerical and Technical Employees in Industry (SIF), and also at certain levels of the SKF management. The main doubts centred on the employee's future: the new work role that would be a consequence of this new technology would be relatively monotonous and require little responsibility. It is well known, both at SKF and in Swedish industry as a whole, that there is great difficulty in attracting younger people for this type of work. Added to this are environmental problems in the form of noise and chemical health hazards. In Sweden the low birth rate has meant that the number of young people looking for jobs will be small during the next 20 years.

In the spring of 1984, the local unions and management made a first attempt to discuss a new development project at the D3 plant. In the summer of 1984 they approached the so-called Development Programme at the Swedish Working Environment Fund for help in organising a project.

In this appication to the Fund, SKF mentioned as the main target the reinforcement of competition with the help of new technology, together with a new work organisation. The parties underlined in the application their wish to create a work organisation adapted to man and the need for all employees to be engaged in the process of change. The application mentioned four development areas, new technology, work organisation, training and work environment.

At the beginning of 1985 the Development Programme accepted both the general approach and the proposed development areas, stressing the importance of broader work roles and increased responsibility for the workers. The project started in the spring of 1985.

4.1.3 Labour relations prior to the start of the project

The employees at the D3 plant are organised in four unions. The majority of blue-collar workers belong to the Swedish Metal Workers' Union, with its own local group organisation in all establishments. The white-collar workers belong to the Swedish Union of Clerical and Technical Employees in Industry (SIF), or the Swedish

Foremen's and Supervisors' Association (SALF), or, like the civil engineers, to the Civil Engineers' Association (CF). As already mentioned, the Swedish unions are strong, with about 70-90 per cent of all employees registered in a union. This applies also to SKF and the D3 plant. As in every large Swedish firm, the unions are well organised, with efficiently functioning channels of communication between the different unions and within each union. The labour relations system at SKF has always been characterised by mutual respect.

4.2 The process

4.2.1 The steering committee

A steering committee was organised with responsibility for targets, overall decisions and follow-up. The committee consisted of representatives of management, the unions (the Metal Workers' Union, SIF and SALF), the Development Programme and a researcher (a research group was contracted by the programme). Management were represented by people both from the D3 plant and from higher up in the hierarchy in order to ensure the necessary power to put the new ideas into practice. The committee consisted of 12 or 13 members.

4.2.2 The project group

Below the steering committee, a project group was created at the local D3 level with representatives of the plant management, the workers and the researchers. This project group was the link between the actual project and the more policy-oriented steering committee. What is significant is that the members of the plant group were on an equal footing regardless of where they worked or who they represented. The group consisted of 11 members, with an SKF official acting as project leader.

The project group was the core group with responsibility for carrying out the project. Through the development groups the participation would widen to include a greater number of employees. In total 40-50 clerical and blue-collar workers worked in these groups for longer or shorter periods of time. An agreement between management and unions gave employees the right to take part in this development work on the same salary as for ordinary work.

4.2.3 The development groups

Four development groups in four different areas were created: new technology, work organisation, training and work environment. Members were chosen on the basis of their knowledge of the area, their willingness to work with the group, and their commitment to the ideas of the project. In addition, there were representatives of the Metal Workers' Union, SIF, management and the researchers, as a sort of resource body.

As well as these development groups there were short-term ad hoc groups formed for specific purposes, for instance to formulate proposals in a particular area.

They usually consisted of representatives of management and employees, plus a researcher.

4.2.4 Setting social, technical and economic targets

The steering committee and project group had as a first aim the formulation of the social, technical and economic goals. These goals were initially formulated in abstract terms, but they were reformulated as concrete and manageable goals so that the development groups could work with them.

Among the social targets were team-based organisation, as much technical and administrative autonomy as possible for the teams, and a great variety of tasks for workers. The team should also have an internal job distribution that allows great flexibility. This means a broader spectrum of qualifications both in the group and in the individual, and as a consequence training of the team members.

The technical and economic goals had been formulated in accordance with the original concept (Concept 1-4), that is, they planned for a more or less automated manufacturing system. The technical goals are confidential and the details are not relevant to this study.

These technical and economic targets were at first sight relatively easy to formulate, but in reality very hard to achieve. From the point of view of management, the targets were formulated in terms of priorities, such as expansion, profit, quality, etc. As regards manufacturing, three targets were formulated: high quality, low cost and flexibility. All this meant that the previously mentioned high-tech concept with FMS cells was still on the agenda. It also meant machines with high precision, NC or CNC machines, software as a help for adjustment and, in the future, computer-aided design and computer-aided manufacturing systems (CAD/CAM).

These targets changed as the project advanced, and here perhaps is the most interesting point in the development of the project. Through unbiased discussions of management's priorities in terms of flexibility, quantity and quality, and their attachment of these to the technical production apparatus, the conclusion reached was that the goals could not be applied to the whole D3 plant. The goals should instead be set from the point of view of the different types of production equipment, where for example FMS cells are only applicable for a certain kind of product, while traditional machines with different work organisation are applicable for other products. The outcome of these goal-oriented discussions was to lay the foundation for work in the different development groups.

4.2.5 The development group reports

Development group – New technology

The task of the work group on new technology was to study the production process in the light of the previously mentioned targets. The group moved slowly and never got round to dealing with the application of new knowledge or technology. Instead it studied problems that arose from the existing technology and proposed several new solutions. The group is still working on proposals relating to the organisation of the maintenance department, an analysis of operations data, a decrease in lead times, etc.

Development group – Work organisation

Taking as a point of departure D3's economic, technical and social goals, the group's task was to work out proposals for an effective work organisation that would allow for a broadening of occupational roles, with increased responsibility, and greater possibilities for personal development at all levels.

Like the technology group, the group was basically forced to deal with the problems of the day. Their most important tasks were a mapping of current organisation, a mapping of planning routines, an analysis of dependence relations, and an analysis of what kinds of tasks the production groups could manage.

Development group – Work environment

This group consisted like the others of both blue-collar workers and clerical staff, but also included a company health representative and a safety engineer. The goals of the group were the following:

- to analyse the reasons for the lack of order and method;
- to survey all workstations from an ergonomic point of view;
- to list workstations that require less strength and mobility;
- to make proposals on how to come to terms with noise, water sprinkle and ergonomic problems;
- to make proposals on the development of new office workstations near the production units.

The group first of all worked on the survey, but they also made proposals for measures to deal with noise, water sprinkle and ergonomic problems, and proposals for new administrative workstations.

Development group – Training

The task of this group was to analyse the need for education among present employees, given two basic targets: that workers in production teams should have a broad knowledge of the work tasks involved, and that the individual should be given the possibility of influencing his/her further development in an expanded work role.

Together with researchers, the group formulated a questionnaire about existing qualifications. As a result, plans for the training of all employees will be drawn up. An educational centre with the same equipment as is used in production is being made ready.

4.2.6 Results so far

As already mentioned, the main target of the management was a high-tech production system with FMS cells. The work done by the different work groups and the researchers showed that this concept did not meet the social, economic and technical targets, and that the expected flexibility in the manufacturing apparatus would not be attained. Instead, it seemed that the existing equipment with minor modifications, and a new work organisation with better training for the personnel, would lead to the same goals.

4.3 Consequences for the workforce of introducing the new technology

4.3.1 Job security

Before the development groups started work, agreement was reached between management and unions that employees should be guaranteed against redundancy or other dismissal: this was known as "job security insurance".

4.3.2 Work organisation

The development group on work organisation produced a proposal for the D3 plant based on semi-autonomous work groups on every production line. The group would be responsible for quality, quantity, delivery times, delivery accuracy and costs. Work roles would be broadened and new tasks introduced at the group level. In the future they would include planning, budgeting, productivity measures, requisition of components, introduction and training of newcomers, grinding, control and assembling – in short, production, and maintenance and repair.

The changes should have been implemented during 1986 but the proposals were put on ice as the management of SKF at that time planned a reorganisation of SKF Norden (the part of SKF that caters for the Scandinavian market). The reorganisation meant that the old sections were made into relatively independent establishments while at the same time the organisation was flattened out, and the number of levels in the existing hierarchy greatly reduced.

The D3 plant got a new director and an intensive discussion took place concerning changes in the internal organisation. This discussion took place alongside the development project but was strongly influenced by the work that had been carried out by researchers and the development groups.

The organisation was in principle similar to that proposed by the development group. It had few levels, and the production teams were to report straight to the chief manager via a production leader. Assisting the production leader there was a "first person" in each team.

The new organisation came into effect at the beginning of 1987 after long and difficult negotiations with the SIF. Problems arose because of the changes in work roles. In general it could be said that the foreman (now production leader) would take on many of the tasks previously done by the manufacturing engineer, while what had been the foreman's job would largely be taken over by the production team and the first man. Through these changes the distinction between what was traditionally clerical work and the work of the blue-collar employees would to a great extent be erased. The foremen would take on more and more of the work tasks previously done by the clerical workers of the SIF. In this way, the stage was set for recruitment conflicts between the different union organisations.

4.3.3 Payment systems

At the start of the development project in 1985, there were many different pay systems in the D3 plant. Some workers were paid an hourly rate, some were paid a piece-rate, and yet others received a monthly salary. After several discussions and negotiations, a new payment system evolved and is now accepted. There is a basic level based on job evaluation, and in addition factors like individual qualifications, group responsibility, performance and productivity are taken into account. The pay system is thus directly related to the competence of both the individual and the group. In addition, the productivity level of the group is taken into account. This system has not yet been put into practice, as more training is required first in order to raise the overall competence level.

4.3.4 Training and retraining

Two of the keywords of the project were education and training. Without adequate training of the workforce, it would be impossible to meet the demands of the new work organisation. As mentioned above, one of the first steps taken was to investigate the qualification structure of the workers. The results showed that many were only trained in one or two tasks. An extensive training programme was therefore started in order to qualify them for more tasks. In the future all workers will be capable of performing every task relevant to their work.

The main training programme lasts six weeks; it includes training on grinding and polishing machines and on assembly machines, and training in statistical process control, quality testing, etc. In addition, there will be training in maintenance and repair work. In the near future, workers will also be trained in planning methods, budgeting, and other parts of the tasks in the new work organisation.

4.4 Effects of the new technology on labour relations

The fact that the labour relations system has always been characterised by mutual respect is of significance for this sort of project, and in spite of temporary quarrels the discussions have been constructive. There was one major incident in 1986, when a decision of importance to the whole project was made at a high level in the SKF hierarchy. They simply wished to stop the whole project. Neither the steering committee nor the project group had received any information. After intervention by the chairman of the Metal Workers' Union, the management took back their decision and apologised to the employees. This may sound idyllic, but relations between management and unions at SKF have always been cooperative. This new project, which in reality implies a rejection of management ideas, did not alter this.

Though relations between unions and management are relatively good, there are problems between different unions, as jurisdictional disputes arise because of the new work organisation and work roles.

5. Evaluation

The three case-studies are simply cases and are therefore not representative of the different industries. Nevertheless, the cases do seem to fall in the mainstream of the Swedish way of handling the introduction of new technology. While relations between employers and unions at enterprise level can be more conflictual than those on the national level, conflicts seldom result from technical changes. Mostly, even in situations of open conflict, unions accept the necessity for constant technological development. We can also distinguish a typical Swedish model for the development of good working conditions, the important components being:

- autonomous goal formulation among the relevant parties within the production sphere (trade unions and employer federations);
- firm external control over working conditions – firstly from the health and safety authorities with their network of local officials;
- state subsidies for work quality improvements;
- tripartite work quality planning processes.

The cases fit relatively well into this model.

The initiative for technological change can come from the unions but usually from the employer. But especially in smaller firms discussions take place between the parties concerning the introduction of new technology at a very early stage. This is what happened in the bank case. The knowledge and experience of the union members were accepted as an important contribution towards successful implementation. In the practical work in joint groups, the unions were able to demand satisfactory solutions to ergonomic or other work environment problems. Here they got powerful support from central unions, laws, and agreements regarding measures to protect safety and health, for example, norms for VDU work. Laws and agreements such as those providing protection against unfair dismissal, which are negotiated on a national level or established by Parliament, are important for work at the local enterprise level.

In the SKF case, the initiative came from the management but the unions were able to alter the proposed technical solutions. Through work in joint development groups, they changed the FMS-based technology to a technically more modest but socially better suited production organisation.

All three case-studies show that training is an integral factor in the introduction of new technology. The unions play an active part both in choosing the right training programmes and in implementing the programmes. This is nothing new: there are national agreements and agreements at industry level to secure the unions' right of co-determination in training measures.

Other important issues are income protection and job security. In neither the newspaper case nor the banking case were there any problems with income protection; the subject was never raised, to our knowledge. At SKF there have been negotiations to avoid lower incomes for workers affected by the new technology or new work organisation.

As regards job security, the solutions followed the typical Swedish attitude. The previously mentioned so-called Swedish model includes an understanding that technical or organisational development should not affect jobs. If this does happen, the management usually agree to provide the workers affected with training or retraining to enable them to transfer to other units. In the case of transfer to other firms, governmental institutes offer the same help.

In none of the three cases were the technical measures and reorganisation of the work process intended to be a purely technical rationalisation which put people out of jobs. Rather, the targets were better quality of product or services and, as a result, greater competitiveness. At the same time these improved products or services opened new market possibilities, which also means new possibilities for employees.

Given a value system in which workplace democracy, democratisation of labour relations, and improvements to the quality of working life are felt to be important issues, the pace of change has decreased and sometimes even come to a stop. After several attempts to introduce change on a vast scale and with far-reaching democratic and work quality aspirations, and after a modernisation of the structural framework of the labour market (of which a modernisation of the legislative framework was an important part), it is gradually being realised today that the traditional strategies have to be rethought.

The traditional way of introducing change, with experimental situations created by researchers and a low level of mobilisation among participants ("expert-created democracy"), has to be replaced by a more participant model. The most central components in this new strategy are the following:

- recognition of locally developed "concepts and theories" and avoidance of general and centre-oriented models;
- recognition of the importance of a local dialogue, including as many people as possible, concerning problems and alternatives and the avoidance of centre-dominated solutions and expert-dominated experimental situations;
- promotion of the possibilities for horizontal learning (learning from each other through common experiences and avoidance of ready-made theoretical knowledge).

This development is just beginning, both in the Swedish Working Environment Fund's Development Programme for New Technology, Working Life and Management and in firms such as Volvo-Kalmar, Saab-Scania and, indeed, SKF.

Notes

[1] S. Edlund and B. Nyström: "Main features of the settlement of labour disputes", in T. Hanami (ed.): *Industrial conflict resolution in market economies* (Netherlands, Deventer, 1987).

[2] L. Ekdahl: *Den fackliga kampens gränser* (Lund, 1988).

6

Technological change and labour relations in the United Kingdom

Roderick Martin in collaboration with Michael Noon***

1. Introduction

Roderick Martin [1]

1.1 National provisions applying to labour-management relations with respect to the introdution of new technology

There are no national-level provisions specifically relating to labour-management relations in respect of the handling of technological change. There is no employment legislation specifically dealing with new technology, and no authoritative national code of practice on new technology. Priority is given to voluntary procedures, arrived at through collective bargaining.

Labour relations issues relating to new technology are covered by legislation relating to other issues, most importantly the Health and Safety at Work Act (1974) and subsequent Code of Practice. The Health and Safety at Work Act provides for the election of safety representatives and the formation of safety committees. The regulations developed under the Act provide, inter alia, for inspection by safety representatives of all substantial changes in methods of working.

The second major relevant national legislative provision relates to redundancies. Employers proposing to reduce the workforce by compulsory redundancy, whether caused by recession or by technological innovation, are obliged to consult with representatives of recognised trade unions. Under section 99 of the Employment Protection Act (1975), the employer must give advance information about the reasons for the dismissals, the numbers and types of workers involved, and the proposed procedures for selecting and carrying out the dismissals. The employer must make reasoned replies to any representations made. In practice, compulsory redundancy, as a result of technological change, especially large-scale compulsory redundancy, is rare, and the provisions of the Act have been most commonly used in connection with dismissals due to falling demand and plant closures.

* Templeton College, Oxford.

** Imperial College of Science and Technology, London.

The third major relevant national provision relates to individual rather than collective labour law, namely the contract of employment. Apart from the requirement upon the employer to provide a written contract of employment, provided for under the Employment Protection (Consolidation) Act (1978), the contract of employment is regulated by an extensive body of common law and statute law. The most important issue relevant to new technology is the extent to which new methods of working constitute such a major change in work as to constitute a new job, and therefore to fall outside the existing contract of employment. The most significant relevant recent common law decision, *Cresswell* v. *Inland Revenue* (1984), involved industrial action by tax officers against the introduction of computers into the Inland Revenue. Tax officers refused to operate the new computerised system, fearing redundancies, and continued to operate the manual system. The Inland Revenue refused to allow the officers to continue using the old system, refused to guarantee no job losses, and refused to pay wages until the officers used the new system. The court held that the employer was justified in his refusal to pay until the officers were prepared to work as required. The judge held that employees had a duty to adapt themselves to “new methods and techniques”, although they could not be asked to learn “esoteric skills”: in-putting and extracting data from computers was a common activity, which an employer could reasonably expect an employee to undertake. (In fact, the tax officers defied the judgement, and remained suspended until an agreement was negotiated with the employer providing for increased flexibility, but with measures to avoid redundancies.)

1.2 Impact of microelectronics on the workforce

Even before the introduction of microelectronics the major intersectoral trend in employment in the United Kingdom was from the manufacturing sector to the services sector: between 1960 and 1973 the proportion of the civilian employed population engaged in manufacturing dropped from 48.8 to 42.6 per cent, and the proportion in services rose from 47 to 54.5 per cent. The trend has continued since 1973, accelerated by the introduction of microelectronics: by 1986 the proportion of the civilian population engaged in manufacturing had dropped to 33 per cent; between 1979 and 1985 the number of jobs in manufacturing dropped by 1,838,000. Part of this decline was due to increased contracting out by the manufacturing sector, involving the transfer of jobs from manufacturing to services, part to declining output, and part to the introduction of labour-saving technologies. Unfortunately it is difficult to establish the contribution of technological change to this trend at the macro level, since several influences are involved, and technology-related employment changes are difficult to disentangle from demand-related employment changes.

According to one estimate, between 1983 and 1985 enterprises that introduced microelectronics reported an increase of 13,000 jobs as a result and a decline of 66,000. The total number of jobs lost directly through microelectronics, both product and process innovations, amounted to 87,000 over the period 1983-85, 1.6 per cent of total employment in manufacturing. Approximately two-thirds of the jobs lost between 1983 and 1985 were on the shop-floor, the remaining third being office jobs, which represents an increase in the contribution of non-manual jobs to the number of jobs lost.

Intersectoral differences do not show a consistent, regular pattern. However, microelectronics has led to an above-average reduction in employment in food and drink, chemicals and metal production, and below average reductions in mechanical engineering, vehicles and textiles. There were, however, major differences in the consequences for job losses resulting from differences in the type of application: where microelectronics was used for the control of individual processes job losses were significantly below the average; where microelectronics was used for the integrated central control of groups of machines or processes job losses were significantly above the average. Overall, microelectronics has not led to a massive loss of jobs, but the rate of loss increased in the mid-1980s, and the new jobs becoming avilable are not in the same places, nor necessarily of the same type, as the jobs that are disappearing. Case-study research suggests that the microelectronics reduces the proportion of process operatives in the manual labour force, and increases the proportion of maintenance workers. However, there are no national survey statistics on the impact of microelectronics on inter-occupational changes in employment.

The level of union membership in Britain dropped from 12,947,000 in 1980 to 10,716,000 in 1985 (including members of non-TUC unions). This drop is partly the result of changes in the structure of employment, particularly the growth of employment, especially part-time employment, in the difficult-to-organise service sector and the decline of employment in highly unionised manufacturing industry. There are no figures for the effect of microelectronics on the level of unionisation. The sharp decline in manufacuring employment, to which new technology has contributed, has resulted in lower levels of unionisation, but it is impossible to specify the contribution of new technology itself to that decline. One point worth noting is that high levels of union organisation have not prevented the loss of jobs – indeed it has been suggested that enterprises with 100 per cent union membership have experienced the least favourable outcomes in terms of job losses.

2. Case-study of a regional newspaper

Michael Noon

2.1 *The context*

2.1.1 The enterprise

The *Sheffield Star*, established in 1887, is an evening newspaper with a circulation of 145,117 from Monday to Friday as a broadsheet, and of 153,606 on Saturday when it appears as a tabloid. It is owned by Sheffield Newspapers Ltd., a subsidiary of the United Kingdom's second largest provincial newspaper publisher, United Newspapers. The *Star*'s principal area of circulation is the City of Sheffield, although it circulates extensively throughout South Yorkshire, extending also into north Derbyshire and Nottinghamshire, with editions for Barnsley, Rotherham, Doncaster, Chesterfield and Worksop. The total population of its circulation area is 1,611,564, with

602,655 households. In addition to the *Star*, Sheffield Newspapers also publish the *Morning Telegraph*, with roughly the same distribution area, and a free weekly newspaper, the *Gazette*, with a circulation in Sheffield of 191,876.

The Sheffield Newspapers establishment sited at York Street, in the heart of the city, comprises a mixture of old and new technology. A full photocomposition production system replaced outdated hot-metal typesetting techniques in 1972, yet the printing presses have not been updated, and the management are presently investigating the possibility of full-colour webb-offset presses.

In 1984 an ATEX computerised direct input system was installed to be used for the composition of all advertising. The intention was for it to become fully operational before a front-end direct input editorial system was added. Thus the *Star* had a fully integrated computerised work process by April 1987. The key feature of these "direct input systems" is that they eliminate duplication of tasks, since the microprocessor technology can capture the primary keystrokes made in the origination area (editorial office) and retain them in memory until the output stage in the pressroom.

The adoption of the ATEX system was not seen as a cost-cutting exercise, according to management: the initial investment was £1.2 million and the rate of return is low, extended over a long payback period. The ATEX system was thus seen as a means of improving productive efficiency and product quality through a more economical use of existing resources.

The management recognised, however, that the technological change would bring with it radical alterations in the work process, and acknowledge the possibility of labour relations problems.

2.1.2 Labour relations prior to the introduction of the new technology

The workforce at the *Star* was organised along the traditional multiple union demarcation and representation lines of British provincial newspapers. The editorial origination area had a predominance of National Union of Journalist (NUJ) members, a few members of the Institute of Journalists (IOJ) (the non-TUC affiliated journalists' union), and several non-unionists, including the editor.

The craft printworkers were organised on a closed shop basis by the National Graphical Association (NGA). Clerical, semi-skilled and manual production workers had the option of joining the Society of Graphical and Allied Trades (SOGAT); however, there was no closed shop and membership was concentrated in production and process, with no union organisation in the advertising sales area.

Labour relations prior to the adoption of the ATEX system were not characterised by any outstanding difficulties. Differences over working conditions and practices were settled through local consultation, and, in the case of the journalists, through a house agreement that was reviewed annually. The introduction of the ATEX system brought about a house technology agreement with the NGA, in which the usage of the direct input advertising system was codified, with a guarantee of job security and limitations on the development of the system, subject to regular review. In this way technology remained part of the traditional enterprise-level collective bargaining system.

A dramatic shift in labour relations occurred in February 1986, not as a result of the new technology, but due to the decision by management to close the *Star*'s sister newspaper, the *Morning Telegraph*. The decision to close the newspaper was, management claimed, purely financial. They argued that the decline in the provincial morning press as a whole, the increasing competition from other news sources and, more specifically, the loss of a valuable source of advertising revenue, the property guide, to an independent rival publisher, made the continuation of the *Telegraph* unfeasible economically.

The importance of the closure is that it soured labour relations, since the management were looking for 240 job losses across the whole workforce. These reductions were based on voluntary redundancy; some staff were transferred to the *Star* on the understanding that they would not be replaced if they left at a future date. Union representatives challenged the management's claim that the closure of the newspaper was a financial decision, arguing that the loss of the property advertising was largely an excuse to rationalise the workforce, provide press time for contract printing, and pave the way for the full direct input system, using ATEX.

The closure of the *Telegraph* created an air of uncertainty and mistrust at the time when Sheffield Newspapers were in between the first and second stages of direct input.

2.2 *The decision-making process*

2.2.1 Decision-making regarding the introduction of the new technology

The decision to invest in the ATEX system was made in May 1983. Prior to this the firm had set up a research unit with a brief to investigate alternative computerised production processes. This unit consisted of the works manager, an overseer, a journalist and the financial controller. Although two of the team were union members, belonging to the NGA and NUJ, they were selected because of their knowledge of computers, not because of their union affiliation. The unions thus had representation on the research unit by default rather than design.

The unit made a series of visits to newspapers abroad – both in Europe and in the United States – in order to assess various systems in operation, isolate those most suitable, and make costings. The findings of the unit were then presented to the board of directors which made the final decision. The ATEX system was deemed the most appropriate by the research unit because it was flexible, user-friendly and powerful; it was also the most expensive. The board of directors accepted the recommendations of the research unit, and invested in the ATEX system. United Newspapers does not have any form of central buying policy, so the Sheffield Newspapers board made an autonomous investment decision.

All computerised production systems introduced in British provincial newspapers have been customised to meet the requirements of the newspaper. In this sense no particular system is ideally suited to any particular newspaper. However, the *raison d'être* behind all new technology is essentially the same: to eliminate duplication of tasks.

There is, however, evidence of some management dissension over the potential of the ATEX system. Although ATEX has the facility to allow page make-up to be carried out on screen, the editor has resisted this (as indeed have the unions), since he is not convinced of the advantages to be gained. Sheffield Newspapers' computer consultant, on the other hand, is continually presenting reports to management on the need to push the technology further. The NUJ is particularly resistant to any change along these lines.

At national level, the NGA has argued for the retention of the double keystroke: the traditional rekeying of all journalistic material. Employers, on the other hand, have argued that technological developments have made this particular process obsolete; they claim that the double keystroke is holding back technology solely so that the NGA can retain control over the production process. This is itself an important issue, since direct input not only threatens the jobs of NGA members, but also threatens to break their hold on the production process. Resistance to technology by the NGA may therefore stem from both economic and political necessity.

The other workers are less threatened and nationally have accepted the new technology as inevitable. The journalists in particular have, however, sought involvement in the process of introduction, arguing that any system involves choices over how it is used, and involvement of the NUJ is a means by which members may both regulate the usage and maximise the benefits for themselves.

2.2.2 *The role of negotiation and consultation*

In general, employers have sought to introduce technology at its full potential, whilst unions have argued for regulation, monitoring and staged introduction. At Sheffield this general pattern is well reflected, with management seeking to have a fully integrated computerised production process within three years of the investment decision being made, the NGA and SOGAT seeking to negotiate some provision within that system for the preservation of traditional tasks and jobs, and the NUJ seeking involvement over the use of direct input in the editorial area.

There is some disagreement between management and unions as to when the workforce representatives became involved in the technology debate. The editor claims that since the unions were involved in the research unit, this occurred at the earliest possible stage, before the decision was made to invest in the ATEX system. The NUJ, however, claim that, by the editor's own admission, their representative on the research unit was there simply for his computer expertise. Thus he was not a union representative, but someone who happened to belong to the union. If he had represented the union he would have been elected by the chapel membership, and would have reported back to them. In addition, when the "representative" left the NUJ to take up a managerial position, there was no official union representative replacement. The NGA acknowledges that it had representation in the early stages, but argue that it was largely on a consultative basis.

Nevertheless, the workforce was not presented with a *fait accompli*, since there was access to information from the research unit, and there were routes to feed in opinions and suggestions. The management claim this was substantial, the unions see it as minimal.

In many respects, whether or not the unions were involved before the final selection is of less importance than would at first appear. The ATEX system is flexible, and its potential far exceeds the present needs of the *Star*. Once the technology had been chosen, therefore, negotiations were by no means complete. There were choices as to how the system should be applied, how to integrate it into existing work processes, and so on. In short the technology itself was not deterministic: its efficient usage largely depended upon workforce compliance and agreement. The management were therefore under some pressure to negotiate rather than consult.

In the second half of 1983, negotiations were conducted with the NGA and SOGAT over the speed of introduction and the problems of overstaffing. Essentially the management set a three-year timetable: the first 12 months being the installation and training period, the following seeing all typesetting and composition using the ATEX system, with direct input for advertising; and the final year seeing the installation of the editorial direct input system and retraining of staff. It was a timetable that the unions could agree to providing that reassurances over job security were given, and that job losses associated with the less labour-intensive technology were absorbed through attrition. Most importantly, the NGA wanted an assurance that there would be no downgrading of their members, and that no one would suffer a loss of earnings through using the new technology. All these points were encapsulated in a written agreement, signed by the management and the NGA.

The second stage of introduction posed more of a problem. As stated above, the closure of the *Morning Telegraph* occurred in February 1986, so when the management approached the unions about the installation of the editorial direct input system in August 1986 they did so in the wake of resentment. In addition, the NUJ and NGA had signed an agreement at national level in October 1985 to negotiate the introduction of direct input technology on a joint basis from that time onwards. At enterprise level this meant that the NUJ and NGA would not negotiate separately with management.

The management at Sheffield agreed to a joint meeting in October 1986 to discuss the timetable for introducing editorial direct input and the consequences for the workforce. Following this meeting, however, there is evidence of a split within management over the validity of joint negotiations. About a week later, the NGA and NUJ received letters from the managing director stating that the management would not entertain the idea of joint negotiations in the future. In effect this blocked any negotiation from that point onwards because the chapels were obliged to follow their national directives to negotiate jointly.

The NUJ full-time officers claim that the reluctance to negotiate jointly was a result of United Newspapers group policy. They argue that every United group title has refused joint discussions in an attempt to split the unions. The Sheffield management claim that the decision was made because they felt no compulsion to conform with an agreement signed between the two unions in which the employers' association, the Newspaper Society (NS), had played no part. Indeed, there was no national agreement between the unions and the Newspaper Society, since early attempts to find common ground over which to negotiate on new technology had fallen through in 1984.

By January still no negotiations had been held, yet the management had unilaterally set a deadline for training to begin on 9 February 1987, and the editorial office had been refitted ready for the installation of video-terminals. At this point a dramatic shift in attitudes to negotiation was sparked off by management. They had already trained IOJ journalists to use the new equipment, and on 5 January 1987 NGA members were handed print-outs produced on video-terminals by IOJ members. The NUJ asked the NGA not to handle any of the IOJ copy. They agreed to this, and the whole of the composing room was locked out. The NUJ stopped work, refused to move, and ignored the threat of an injunction. By 4 p.m., however, the NGA had negotiated their way back into the building on an instruction from the NGA head office. This highlighted the important fact that the NGA were unlikely to risk any action that would compromise their bargaining position over redundancy payments. It also showed the NUJ that they could not rely on the NGA for supportive action in the case of a dispute.

The NGA and NUJ now agreed to abandon their insistance on joint negotiations, provided that the management allowed observers to sit in on the other union's negotiations, and that these negotiations would run in parallel so that each union would reach agreement at approximately the same time. The management agreed, and negotiations began on 2 February 1987, a week before the management's training deadline.

The two unions had insisted on joint talks largely in order to avoid the demarcation problems that editorial direct input would bring about. In several British provincial newspapers, the unions had come into conflict over management-NGA deals that allowed NGA members to transfer into editorial jobs by performing technical subbing on screen. The NGA claimed that they were simply "following the job" and that their tasks had been transferred. The NUJ did not object to this in principle, but to the fact that these transferees were still represented by the NGA for collective and individual bargaining purposes, and that they retained their NGA pay levels, which created anomalies such as NGA transferees being trained in editorial subbing tasks by skilled journalists who were on lower wages.

In October 1985, the unions finally agreed that any transferees would become joint members of the NUJ and NGA. Furthermore, it was agreed that neither union would be sole signatory to a deal without the consent of the other. That is why the negotiations over editorial direct input could not take place until the compromise of using observers had been found.

Three sets of negotiations then took place. The least problematic were those with SOGAT, largely because of the union's weak position at Sheffield. The discussions concerned voluntary redundancy and transfer. Offers of redundancy were made and one SOGAT member took up the offer.

The negotiations with the NGA were also concerned with job losses and transfer of work. The management offered substantial redundancy payments to NGA members in an attempt to reduce the workforce. There was discussion on the number of redundancies, but not on the payments offered, and a figure of 21 was eventually reached. Discussion also took place on transfer, and the management agreed to offer the opportunity for four transfers into the editorial area. The negotiations went fairly smoothly, according to both sides, largely because NGA members were willing to accept the terms of voluntary redundancy on offer.

The negotiations with the NUJ were more problematic since the deadline for training would be reached in five days. The NUJ had a national policy of not touching any equipment until an agreement had been signed, but the management could not accept this; they claimed that even if the journalists were trained they would still have the ultimate sanction of refusing to use the equipment, so they were not prepared to push back the deadline for training.

Negotiations took place over ergonomics, health and safety, and job security. It was agreed that there should be no job losses (voluntary or otherwise), full health and safety provision in line with TUC guidelines, new office furniture and workstation facilities as recommended by the NUJ head office, and provision for pregnant women to transfer from VDU work. There were, however, two sticking points: *screen breaks* and *technology payments*.

On the first of these points the NUJ argued that independent research had shown that it was vital to have regular off-screen breaks in order to reduce stress and eye-strain and lower the risk of other recognised industrial injuries such as tenosynovitis. They were requesting statutory screen breaks of ten minutes every hour. The management argued that the nature of journalistic work meant that natural breaks occurred, so statutory breaks were unnecessary, but they were prepared to allow breaks every two hours.

The second problem area was how much money the journalists should be paid for retraining and using the new equipment. Management were offering across-the-board payments for reporters and subs, with a lower rate for photographers, since they would not be required to use the equipment. The NUJ would not accept this, for two reasons: first, they claimed that photographers and journalists should be paid the same since they belonged to the same union, worked in the same editorial area, and would be expected to conform to new working practices associated with the new technology, for example the moving of the copy deadline. Second, the NUJ argued that in any case the payments were not sufficient: compared with other newspapers belonging to United, their basic wages were poor. By the fifth and final day neither side was prepared to move. The management then produced an improved offer for the journalists, although not for the photographers. This was put before a chapel meeting that evening, and rejected by 93 votes to six after a three-hour debate.

On 9 February, when training was due to start, the two sides negotiated a 24-hour cease-fire and tried to settle their differences. A compromise over screen breaks was reached, with both sides agreeing to a ten-minute break after 90 minutes, but the technology payments were not resolved. The NUJ was then given an ultimatum to commence training at 2 p.m. the following day or be in breach of contract. The consequence was a dispute.

The first five people selected for training were the two assistant editors, the features editor, the chief sub and the deputy chief sub. The NUJ claims that these people were selected because they were most likely to disobey the union instruction not to begin training until agreement had been reached. The management claim that they were selected because they were the most senior members of staff and would be overseeing the whole process. The two assistant editors resigned from the union and began training, but the other three refused; they were given 48 hours to change their minds or face dismissal.

The following day NUJ members arrived for work and were individually asked by the editor whether they were prepared to train. Each refused to do so without an agreement, and so was not allowed into the building. The union claims this was a lock-out, while the management argued that a refusal to work normally, including training, constituted a strike.

The dispute lasted a month, with none of the 100 NUJ members working throughout that period. An approach was made to ACAS by the NUJ Father of Chapel (FOC), the name given to shop stewards in the newspaper industry, but this was rejected by the management. Eventually, realising that the management were unlikely to move, the union's head office advised a return to work, accepting the management's final offer over new technology.

The dispute was seen by the chapel as a climbdown, achieving very little. There was also some resentment of other trade unionists who had continually crossed picket lines, although the chapel realised that the NGA would have been unlikely to take supportive action and risk their favourable redundancy terms.

Essentially the dispute was over the financial benefits to be gained from the introduction of the new technology. The journalists felt their potential new skills were being under valued, while the management saw the dispute as arising from greed. They felt the journalists were overpricing themselves and that the benefits accrued were in any case more than financial since the workers would have a new, more comfortable working environment and would acquire new skills that would increase their value in the labour market, open up job opportunities in the United Kingdom and abroad, and enhance their promotion chances.

The dispute ended on 10 March 1987, and by 6 April the *Star* was using a total direct input production system, two months later than the management had originally planned.

2.3 Consequences for the workforce of introducing the new technology

2.3.1 Job security

The full effect of the new technology on job losses is hard to measure, for various reasons. First, the management negotiated a no-replacement policy, claiming that this would provide a means of rationalising the workforce in the early stages of introducing ATEX. Second, the closure of the *Morning Telegraph* led to the loss of 240 jobs, some of which involved working for both newspapers, as they shared the same premises, plant and production facilities. The unions claim that some job losses might have been attributed to the closure of the *Telegraph* when in fact they were part of a wider rationalisation process due to ATEX. This is denied by the management, who argue that in fact they kept many people at work unnecessarily, though agreeing with the unions that they would not be replaced once they left the firm.

The second stage of direct input brought a cut of 21 jobs from the composing room, achieved through voluntary redundancy. Had this figure not been reached, the management would have resorted to compulsory redundancy. For those employees who left the firm, the management paid a lump-sum redundancy payment based on number of years' service and age. Future rationalisation would be based on transfer and non-replacement of staff.

The management offered full job security in the editorial area with no redundancies. In addition, they offered four job opportunities in the editorial room for employees wishing to transfer from the composing room, but these jobs were not to be filled by outside workers, if existing employees did not wish to take up the offer, or if the applicants were regarded by management as unsuitable.

2.3.2 *Work organisation and working conditions*

Substantial changes in working practices were entailed by the second-stage move to editorial direct input. For production workers, it meant abandoning the process of rekeying all text produced in the editorial area. This task disappeared completely, although the task of keying in typesetting and format instructions was transferred to the editorial area. In addition, reading tasks were reduced and the responsibility of ensuring clean copy was transferred from the production room to the reporters and sub-editors.

An example of the way in which the technology has reduced the number of individual tasks can be seen by looking at the making up of type for an advertisement. Before direct input it involved three processes with different workers: a "mark-up" person to select typeface, size and so forth, a keyboard operator to transfer this to the photosetting equipment, and a "paste-up" person to compose the advertisement using photoset type. Direct input allows the whole process to be performed by one operator, viewed on screen, sent back to the main computer, and then output by the phototypesetter.

Primary keying of externally originated material, such as television copy and freelance material, would be directly input by production workers, and recalled in the editorial area; however, wire copy capable of being directly input into the system would not be retyped. Furthermore, there would be no retyping of copy from district offices or from journalists at outside locations such as court, sporting events and so forth. These would be written into portable lap-top computers and transferred directly into the system via telephone lines, using modem units. Journalists had begun to use these in 1984, but all the text was produced as hard copy on a printer and retyped. From April 1987 this retyping stage was eliminated.

In order for the subs to cope with the new work process the management suggested that the editorial deadline be brought forward to prevent a backlog occurring at the subbing stage. This was reluctantly agreed to by the reporters, although the deadline was returned to its original time after three months when it was felt that the learning process had been completed.

The reporters and the subs basically consider themselves to have acquired new skills. The subs in particular have acquired some of the production skills that were formerly the domain of the NGA workers.

While the general trend has been to increase skills in the editorial area, the opposite is the case in the production area. The compositors are expected to perform a greater variety of tasks, but the information-processing capacity of the computer has speeded up the tasks and simplified the whole work process. Management argue that the simplification is not a restriction, but enhances the potential to produce a better product. The assumption here, of course, is that an enhancement of the product means that the worker derives greater satisfaction. This need not necessarily be so if the task becomes more monotonous and less operator-determined, with the worker being peripheral to the computer. The compositors are using keyboard skills to enhance the product, but not the manual skills that might enhance the work satisfaction.

Management and unions have agreed that all further stages of technical change would have to be negotiated prior to introduction. There is also provision for quarterly review of existing work practices to monitor the progress of changes and provide means by which grievances and problems can be aired.

As far as working conditions are concerned, the main area of concern was with the VDU workstation. Management consulted with the unions, both at the initial stage of introduction in the production area, and then in the editorial area, about suitable office furniture, seating arrangements, lighting, humidity and noise levels. There was a mutual agreement that the guidelines laid down by the TUC health and safety committee should be adhered to, so work areas were refitted accordingly.

A problematic area was health and safety, the management arguing that the VDUs posed no threat to the health of the operators while, as already discussed, the NUJ insisted that operators should be given statutory breaks from the screen, and that operators who were pregnant or had permanent health conditions preventing them using the VDU should be given the option of transferring to similar off-screen work, losing no money or status. The management accepted statutory breaks in the production area as part of the normal work process, but were reluctant to extend this to the origination area. Eventually they agreed to do so.

Management also agreed to pay the cost of eye-tests for all workers who transferred to VDUs, and to reimburse any worker who had to wear glasses or already wore them and needed a change of prescription as a result of the change in work task.

2.3.3 Payment systems and income protection

The editorial workers have the same pay system as before, based on a house agreement and banded rates of pay according to seniority. However, all the basic rates have gone up, as outlined above. The journalists do not receive productivity bonuses or overtime payments, so the new technology has had little effect on their pay system.

The printworkers' pay system also remained unchanged, although the new technology is thought likely to bring about an eventual reduction in overtime by speeding up the production process and making it less labour-intensive. As yet there has been no change, and management have argued that overtime is likely to be sustained through the system's potential to print more titles and handle a greater output.

Management agreed that there should be no downgrading of work, or loss of basic earnings, but that where a new work task was established, a new rate of earnings would be deemed appropriate, and that any new employees would operate the technology at that rate, irrespective of the rate paid to transferees to that particular work task. Where existing tasks were enhanced by the technology, as in the case of editorial tasks, it was agreed that the new basic rates of pay should reflect this and be applicable to new employees as well as those who had retrained. In the case of transferees to editorial from composing tasks, it was agreed that their earnings should be protected, but that parity with colleagues doing the same task should be the principle. Any excess income level would therefore be frozen until transferees and non-transferees were earning the same amount.

2.3.4 Training and retraining

The ATEX system brought with it the need for considerable retraining of staff in all departments. The management agreed to a full training programme to meet the new requirements both in production and in editorial. The most substantial retraining was for sub-editors, who were expected to convert their previous paper sub-editing tasks on to screen, with new instructions and formats to be learned. The reporters had to acquire word-processing skills, but these were very similar to the typing skills already held. Production workers had to learn the use of the more sophisticated typesetting laser equipment compatible with ATEX and the more extensive use of on-screen advertisement design. In addition, as already mentioned, the management offered four editorial positions for NGA members who wished to transfer from the composing room. Transferees would enter the editorial department as adult trainees and complete the appropriate training course. Out of the original five inquiries, only one person was in fact selected – a 45-year-old compositor.

All the training programmes were designed by the management, in consultation with the suppliers of the ATEX system. Training sessions were conducted in small groups on a "train and replace" basis, a new person starting training once a sufficient level of competence had been achieved by another trainee. The last group to retrain were the reporters – the photographers were not required to use any of the new technology and consequently have been given no training.

2.4 Effects of the new technology on labour relations

2.4.1 Effects on the structure of the workforce

The reskilling of journalists and sub-editors undoubtedly increases their value in the labour market, especially for jobs abroad where direct input is used extensively. Clearly the ability to use direct input will have a declining market value as more journalists are reskilled; within the next few years the "skill" will become a necessary and integral part of the job as nearly all provincial newspapers convert to direct input.

Within the production area, the move away from single-task work to multi-task production work might increase an individual worker's mobility. However, decisions over when and where workers should be redeployed is likely to remain a management decision. Indeed, they perceive one of the virtues of the system to be greater workforce flexibility in the production area.

2.4.2 Effects on the unions

The loss of jobs in the composing room at the *Star* has meant that the closed shop NGA has lost membership. This is recognised by the union as an inevitable consequence of direct input technology, and attempts nationally to hold back the technology have been as much concerned with the potential loss of control over the production process as with the preservation of jobs *per se*. At the *Star* the relative strengths of the NGA and SOGAT chapels deteriorated as technological change became inevitable. Thus, by the time the editorial direct input system was introduced the unions, and the NGA in particular, were more concerned with negotiating redundancy payments than with supporting the NUJ in their dispute. Indeed, when the NGA did take industrial action, they were instructed by their head office to go back to work since a dispute was likely to threaten the redundancy negotiations.

The NUJ differs somewhat in that the *Star* chapel clearly perceived it could exert greater industrial muscle due to the new technology, and therefore entered into a dispute to ensure a maximisation of benefits. In terms of numbers, the NUJ had 104 members before the technology and 100 afterwards because four members resigned over the decision to enter into a dispute.

There was an increased interest in the union immediately prior to the dispute; this interest, according to the FOC, has diminished slightly but still remains high. National officers of the union argued that the new technology has increased the funtions of the union since there is now a need for greater negotiating and monitoring of the work process.

An importance change in demaraction patterns has been brought about by direct input. The nationally negotiated NUJ-NGA joint agreement acknowledged the journalists could perform certain production tasks that were formerly the domain of compositors. Additionally, it laid down the principle of federated house chapels and negotiating committees, although this has not occurred at the *Star*. In fact, there is evidence to suggest that the unions feel more alienated from each other than before the new technology was introduced. The NUJ FOC claims that there is a residual bitterness against the NGA, who did nothing to help the NUJ chapel. This hostility does, however, predate the dispute, since there was an active group within the *Star*'s NUJ chapel arguing that they should ignore the head office directive to push for joint negotiations alongside the NGA, and instead negotiate alone. As it transpired, separate negotiations did occur, but this was at the insistence of management not the unions.

A further worry of the NUJ nationally is that direct input brings with it the change for management to employ an increasing number of freelancers, working from home. If this occurs it might bring about a reduction in membership. At the *Star* there is provision for using outside contributed copy, but the chapel was keen to ensure tht the scale of this did not escalate and undermine staff copy; thus any change has to be negotiated.

In general there is a feeling amongst the unions that non-union labour could be more easily used to produce the newspaper than before. The NUJ FOC claims that the management trained secretarial staff to input using the ATEX system, normally not part of their work tasks, and that these members of staff could be used to undermine union effectiveness in times of dispute. The unions recognise that even if the production workers had supported the NUJ in their dispute, the newspaper would still have been produced. This has been confirmed by the editor, who claims that there is a contingency plan enabling the *Star* to be produced on the assumption that no trade unionist in the building is working. This is a dramatic shift from the past when, although NUJ action has rarely stopped production of a newspaper, the NGA have always been able to prevent publication and have wielded considerable industrial muscle.

2.4.3 Effects on management

Technology itself has not brought about any change in the structure of management, although the structure has changed over the past three years as a result of a new managing director going into Sheffield Newspapers.

The editor claims the Sheffield Newspapers management has always had considerable autonomy from the United Newspapers management. The unions question this: the NUJ in particular has argued that there was a discernible United Newspapers policy over the rejection of joint negotiations. They claim that the local management were initially prepared to hold joint discussions, but were overruled by the central management at United Newspapers.

2.4.4 Patterns of negotiation and consultation

There is some disagreement between management and unions over the state of management-employee relations at the newspaper. Both sides recognise that the closure of the *Morning Telegraph* embittered relations, but the management deny that this had any spillover effect into the technology negotiations. Management claim that all employees were willing to accept the technology. They argue that the inevitable claim and counter-claim discussions during the technology negotiations were in line with traditional bargaining practice and there was no animosity. Although this might be true of the manual unions, it was certainly not the case with the NUJ chapel.

The FOC argues that there is at present contempt and outright hatred for the management as a direct result of the technology negotiations. He points out that the last thing the chapel did before returning to work on 10 March 1987 was to pass a vote of no confidence in the editor, which was carried 84 votes to 0. Furthermore, he claims that this ill feeling has manifested itself in the work environment in two ways. First, whereas before direct input journalists at the *Star* would work long hours and chase stories, they now worked to their hours and noted every extra minute for days off in lieu. Second, since the introduction of direct input 12 of the journalists have left the firm, the FOC included.

This, he argues, is a result of a general distrust of management and the erosion of any feeling of loyalty for the firm during the month-long dispute. The editor, on the other hand, argues that these journalists might have left in any case; if anything, the new technology has increased their mobility and value in the labour market, thereby accelerating the pace of staffing changes.

In theory the new technology should have improved the potential for negotiation in the future and facilitated more contact between management and unions, since the agreements signed between management and unions allow for regular review of the operation of the ATEX system and new working practices associated with it. This should pre-empt conflict, first by providing a system for monitoring and discussing problems in their infancy, and, second, by involving the unions in any proposals to extend the technology at the decision-making stage, rather than the implementation stage.

In practice, the quarterly meetings have so far brought to light only minor issues. One of the intentions of the NGA-NUJ national agreement was to provide the means of joint consultation with management over the use of the technology, but clearly this has not happened at the *Star*. Management say the workforce has adapted very quickly to the technological change. They claim that they have complied with the technology agreements to the letter, and that any minor practical difficulties have been sorted out, either directly or through the review meetings.

2.4.5 Conflict

In the production area the effect of new technology has largely been to dampen down conflict. As can be seen from the account of the one-day dispute in subsection 2.2, the direct input system brought with it an uncertainty over jobs that softened the militancy of the NGA chapel when faced with the prospect that if they took strike action the management would be able to sack workers for breach of contract and therefore not have to negotiate redundancy. In short, the technology broke the industrial muscle of the NGA by making many of their traditional skills obsolete.

In contrast, the journalists have if anything become more militant. Apart from the initial dispute over technology payments, the NUJ chapel became involved in a further dispute, this time over the house agreement, six weeks later, when the chapel voted to work to rule. Members soon became disillusioned, however, so the action collapsed. Traditionally, journalists have never been particularly militant, but their militancy at the *Star* might indicate a general feeling that they are now in a position of power they have never previously enjoyed.

2.5 Evaluation

In spite of the difficulties in introducing the new technology, the outcome has on the whole been favourable to both management and workforce. This outcome does, however, mask many of the underlying labour relations problems. Almost from the time of announcing the purchase of the ATEX system to its full operation, management and workers came into direct conflict. Faced with the prospect of managerial decisions over technology making obsolete most of their jobs, the production workers acknowledged that the technology was essential for the future of the firm, and agreed to cooperative providing management took a gradual approach to its introduction and were willing to negotiate transfer, voluntary redundancy and a policy of attrition. By the time the second stage of introduction was proposed, after the decision to close the *Morning Telegraph*, the production workers, although already weakened, nevertheless felt strong enough to take industrial action over the setting of non-NUJ VDU-produced copy. However, in the fact of the management response – a lockout – they backed down.

A similar policy was adopted towards the journalists. The management refused to negotiate jointly, but set a deadline for training, knowing that the NUJ would refuse to train until an agreement had been reached. Finally they trained non-union journalists and told them to use the VDUs – which worried many union members. As a result the management got their own way and the unions agreed to separate negotiations.

Negotiations with the production workers went smoothly because the subjects under negotiation had been discussed previously and the agenda set by management was one with which the NGA could agree. The parallel discussion with the NUJ brought about conflict on a series of points, the impasse over technology payments and the imposed training deadline forcing the chapel into industrial action. The damage this conflict caused to relations between management and editorial workers is apparent in the bitter terms in which many journalists refer to management. On the other hand, management do not perceive any change in the attitude of the editorial workforce, arguing that relations with the unions have always been good, and remain so.

3. Case-study of a major national clearing bank

Roderick Martin

This case-study is of an ongoing project in a major national clearing bank, the Branch Information Technology project, due for completion in 1989. The particular case-study was chosen for three major reasons. First, it is a very large-scale project, costing over £600 million, and of central importance to the bank's future. Second, it is in the information technology area, which is likely to be a major source of competitive advantage for banks in the 1990s. Third, studying an ongoing situation avoids the danger of rewriting history through retrospection. The disadvantage of studying an ongoing project is obviously that the story is incomplete, and final assessment impossible. However, I hope the advantages outweigh this disadvantage.

3.1 The context

3.1.1 The enterprise

The bank is one of the four major British clearing banks which dominate retail banking in Britain. It has 2,200 branches, with an overall total of over 56,900 employees in 1986. In addition to its retail branch network, the bank has an investment and securities branch, and an international division. Although not the largest clearing bank, in the early 1980s it was the most successful financially, partly because it had fewer losses overseas than its major rivals. More recently the financial position has deteriorated because of the Latin American debt crisis. In 1986 the bank achieved a profit of £700 million before tax, £470 million after tax; the major part of the post-tax profits, £267 million, or 57 per cent of the total, was achieved by the UK Retail Banking sector of the bank – the sector involved in the Branch Information Technology project.

3.1.2 Reasons for introducing the new technology

Since 1970 there have been major changes in the banking industry, which have involved all the clearing banks. Competition between the banks, and between banks and other financial services institutions, has increased, especially with a broadening of the scope of the activities of building societies. Foreign banks have entered the British market, either directly or through the purchase of British banks. Deregulation and the "Big Bang" have transformed the City of London, and clearing banks have begun to play a major role in the Stock Exchange. In retail banking itself, the volume and variety of financial transactions has increased.

These major changes in the banking and financial services sector have increased the pressure for the adoption of new technologies, for several reasons. First, there is the need to cope with the increased volume of financial transactions. Second, costs need to be reduced to maintain a competitive position vis-à-vis other financial institutions, especially building societies, as regards personal savings. Third, increased sophistication is needed in the handling of the bank's treasury role, especially as major customers themselves have acquired increased skill in the management of their own funds. Finally, bank managers have shared with other managers the desire to introduce more sophisticated management information systems.

There have been four major strands in the development of new technology in the British retail banking sector, in which the case-study bank has played a major role. The major technological changes have been the introduction of automated cheque-clearing facilities, including the most recent development, the Clearing House Automated Payments System (CHAPS), with same-day settlement of cheques for £10,000 plus via the Bank of England; the development of Electronic Funds Transfer at Point of Sale (EFT/POS), whereby customers' accounts are electronically debited when purchases are made at participating stores; the use of automated telling machines (ATMs), whereby customers can undertake routine transactions (for example, cash withdrawal) themselves – hole-in-the-wall banking; the use of "back-office" automation to increase the speed of information-processing and to reduce the amount of paperwork. The Branch Information Technology (BIT) project, the subject of this case-study, is a major example of the fourth strand.

The introduction of BIT is being planned as a phased programme, running from May 1987 until September 1989. Investigation of branch operations in 1985 led the bank to conclude that major changes were required: other banks were increasing their levels of automation, and the existing computer system was incapable of coping with expanded requirements. The technology chosen was a large, mainframe system with remote on-line access, with processors, video-terminals, keyboards and printers in each branch; there will be at least two keyboards for every three members of staff. The system includes a new telecommunications network, with multiple routes between each branch and the central computer: this will increase the reliability of communication between branch and centre. The new system involves 27,500 new pieces of equipment, at a cost of almost £600 million, and the integration of 6,000 existing machines into the new system. The project involves two parallel strands, equipment installation and applications: initially, the new equipment will be used for only a small number of applications.

The overall strategic objective of the BIT project is to enhance the bank's performance in the sector from which it derives the greater part of its profits, UK retail banking. According to the bank's chief executive, "Our aim is to differentiate ourselves from our competitors by better products, innovation, and, above all, a high level of service." The bank's strategy involves market segmentation, concentrating effort where it believes it possesses competitive advantage. To help achieve selectivity the bank has divided responsibility for services according to the size of account: branch managers are to concentrate on personal and small business customers, while medium-sized customers are to be served by a newly formed commercial services division. Such a strategy requires an effective information system.

The bank also wishes to reduce operating costs through the use of new technology. In 1986, staff represented 63 per cent of operating costs; staff costs rose by 9 per cent between 1985 and 1986, significantly more than premises and equipment (which rose by 3 per cent) but less than miscellaneous expenditures, including advertising. The bank's performance on operating costs improved in the 1980s, especially in the UK retail banking sector: in 1982 the costs:income ratio was 73:100, in 1985 64:100. New technology could make a major contribution to reducing operating costs in the future by reducing the number of employees (or holding back the rate of increase) and by reducing the cost of premises.

The major operating objective of the new system is to reduce the amount of paper generated in the bank by automating information currently held in paper form (for example, the daily record of balances). This will reduce staffing and space requirements, and make the information more manipulable. It will be possible to group information on a customer – not simply an account – basis, and to have easy access to information from branches other than the branch at which the account is held. Eventually customers will be able to have direct access to account information via visual display screens. In short, BIT is intended to lower operating costs by reducing the amount of clerical labour and by reducing premises costs, to increase the sophistication of marketing activities, to increase the range of financial services made available to customers, and to improve management information.

At the time of writing the BIT project is of course incomplete: the installation of equipment and the training of personnel is under way. The first stage of the project was due for completion by the end of September 1987. This involved the transfer of existing facilities to the new equipment (for example, ledger details, yesterday's entries and charging statistics); the development of the facility to transfer reports electronically from the computer centre to any outlet, at a limited number of branches initially; and the completion of file maintenance/input via the new equipment.

3.1.3 *Labour relations prior to the introduction of the new technology*

At the time of launching BIT, labour relations were carried out at two levels, at industry level and at enterprise level. The issues negotiated at industry level comprised terms and conditions of service (including holidays and London allowance) and the minimum and maximum points on the pay scale for clerical grades 1-4 and some supervisory staff. Negotiations were carried out between the Federation of London Clearing Bank Employers, on the one hand, and either the Banking, Insurance and Finance Union (BIFU) or the Clearing Bank Union (CBU), on the other. Issues not specifically allocated to the industry level were negotiated at enterprise level. New technology was an enterprise-level issue.

The situation was transformed by the withdrawal of the National Westminster Bank from the Federation in 1987 (while the case-study was being prepared), leading to the collapse of the Federation, at least in the short term. This was due to differences between the banks over the annual pay offer and the level of increase in the London allowance. At the time of writing the pattern of labour relations in the industry is changing rapidly. It is likely that in future all negotiations will be conducted at enterprise level. Since the major clearing banks are highly centralised organisations, there are no negotiations within the enterprise at regional or individual establishment level, although there are separate "sectional" negotiations for specific groups of staff (for example, staff working in data processing, financial services and trusts).

Union structure in the enterprise is simple, but problematic. Three unions have been involved in the industry in general: the Association of Scientific, Technical and Managerial Staffs (ASTMS), the Banking, Insurance and Finance Union (BIFU), and the Clearing Bank Union (CBU). However, the CBU is in the process of dissolution, its constituent unions, including the staff union in the case-study bank, operating independently of each other. The largest union in the bank is the staff union, with approximately 26,000 members. Although it evolved out of a staff association, initially largely sponsored by management, it is now completely independent of management. The union was formerly a constituent part of the CBU, but has always had its own officials, staff and headquarters. Unusually for the United Kingdom, it is an enterprise union. It is not affiliated to the TUC. Negotiations at industry level were conducted by the CBU, the negotiating committee including representatives from the staff union; negotiations at the enterprise level are conducted directly by the staff union. The second largest union in the bank, but a larger union overall, is BIFU, which recruits workers throughout the banking and financial services sector, with particular strengths in one of the other major banks. At the time of the case-study research it had nearly 162,000 members, including just over 13,000 in the case-study bank. BIFU is affiliated to the TUC. The third union involved in the retail banking sector, the ASTMS, has no members in the bank.

A major factor in labour relations in the case-study bank is competition between BIFU and the staff union. The rivalry is intense, being based on ideology as well as institutional interest: BIFU does not regard the staff union as a genuine union, mainly because of its non-affilitation to the TUC, and its too close relationship with bank management; the staff union does not regard BIFU as an effective agent for employees within the bank. The staff union bases its views on the special employment relationship in banking: lifetime employment, internal promotion, limited recruitment from the external labour market, and basic long-term identity of interests with the employer.

Labour relations in the British banking industry have traditionally been non-conflictual. The case-study bank has never experienced a national strike. The bank has traditionally adopted a paternalistic policy, providing employment security and good if not spectacular earnings. Traditionally, promotion prospects have been good for employees who pass the Institute of Banking examinations, at least to the so-called appointed officer grades. Promotion prospects for male employees have been sustained by the high turnover of female clerical employees, relatively few of whom are promoted even to the appointed office grades, although the bank has an equal opportunities committee. Interrupted employment histories and the difficulties of moving between branches help to explain the relatively small numbers, although the situation is changing, especially in London and the south-east. At the time the case-study research was begun the enterprise was experiencing an overtime ban in connection with the annual pay negotitions, both BIFU and the staff union rejecting the 5 per cent offered (and subsequently imposed) by the Federation of London Clearing Bank Employers. Following the collapse of the Federation, individual bank managements improved their offers, which were accepted by both unions.

Overall, labour relations in the bank are in a process of transition, with the level of conflict increasing, but from a low based. This increasing militancy in part reflects the strains on the traditional employment relationship and paternalistic management system.

3.1.4 General characteristics of employment relationships

British clearing banks have traditionally adopted a paternalistic employment policy. There is a high level of job security, with no history of compulsory redundancy; in only a relatively small number of cases has voluntary redundancy been used (including the case-study bank): optional early retirement is more common. Voluntary redundancies have occurred amongst technical and service staff, and head-office departmental staff. The cause has been internal reorganisation, not new technology. Traditionally, there has been a low level of labour turnover amongst male staff, except in a small number of technical occupations. This has been partly due to the bank's own employment policies, and partly to a "no poaching" agreement between the major banks.

The level of labour turnover amongst female employees is higher, reflecting the bank's policy of recruiting directly from school, women characteristically ceasing full-time employment in their late twenties to bring up a family. In the past, relatively few female employees have returned to banking after rearing children, since the bank has traditionally had very few part-time employees: however, the number of part-time employees in banking is now increasing, many part-timers being previous employees returning to employment.

The bank operates a national employment policy, recruting to the bank rather than to a specific branch. Employees are therefore expected to accept redeployment between branches, and at managerial level between regions. Acceptance of redeployment is obviously a necessary precondition for employees hoping for promotion: vacancies are not usually filled by promotion within the individual branch, but on the basis of grade, experience and performance assessment at regional level. One consequence of this need for mobility is the difficulty female employees have in obtaining promotion, since women have difficulty in persuading their spouses to move. The proportion of women in managerial and supervisory grades is therefore much lower than the proportion of women in the bank's employment.

Since 1971 the bank has operated a national job evaluation scheme, common to all major clearing banks, for clerical staff. Jobs are classified into four grades, each grade with its appropriate scale. Each bank has its own scheme of job evaluation for appointed and managerial staffs. The decision whether a particular job is classified as appointed or clerical is a managerial one, but within each category, and within each grade, the ranking is agreed with the union. The job evaluation scheme replaced an earlier age/person-related scheme. One of the issues raised by new technology is the need to revise the job evaluation scheme to reflect the major changes that have taken place in banking work since 1970.

3.2 The decision-making process

3.2.1 Decision-making regarding the introduction of the new technology

As a result of recent reorganisation, the case-study bank is now organised into five major groupings, with a small corporate head office exercising overall responsibility for the bank's corporate strategy, including the emphasis to be given to different areas of activity. The major responsibility for deciding upon the BIT project lay with one of the five groupings, United Kingdom Retail Banking (UKRB), and the project was specifically seen as a UKRB project, unlike major computer projects in the past. This allocation of responsibility for new technology was part of an overall strategy of delegating downwards whenever possible, each grouping exercising individual financial responsibility: annual financial results are presented at the individual group level.

As with other aspects of the bank's activities, overall responsibility for research on new technology belongs at the corporate level. However, each group is able to develop its own proposals for new technology, based on its perception of its technical requirements: hence UKRB carried out the research on and selection of equipment for BIT. Within UKRB the extent to which individual units are allowed autonomy on technical issues depends on the degree of integration of the particular activities with the bank's activities as a whole: if the activity is self-contained, systems can be adopted that are not necessarily compatible with overall systems requirements. Responsibility for deciding upon the issue lies with the general manager responsible for systems development within UKRB.

The scale and cost of the BIT project require a very large project team. This team is led by a member of the central management group within UKRB. The team itself consists of four senior managers and a team of 30 drawn from the branch banking sector. One member of the team is designated to cover training, communications, personnel aspects of data loading and the implications of BIT for job descriptions. The project team is specifically concerned to ensure that the technology adopted should be "user-friendly". This objective influenced the composition of the project team, which includes managers experienced in branch banking. The project team is composed solely of managers: there is no union or employee representation.

3.2.2 The role of negotiation and consultation

Bank management regard technological change as an integral part of management responsibility. There were therefore no negotiations over the introduction of BIT, the choice of technologies, the speed of technological change, the consequences for employment, job classification, pay systems or other working conditions, the distribution of benefits arising from the new technology, or compensation for any adverse effects arising from the new technology. The general bank union, BIFU, pressed for the negotiation of a new technology agreement in the bank, but without success.

The staff union did not believe that a specific new technology agreement was necessary, but instead preferred new technology to be treated as part of an overall "participation agreement" to deal with all issues of change, including new technology.

One potential area within which bargaining might have arisen was the job evaluation scheme. The job evaluation scheme was negotiated between the Federation of London Clearing Bank Employers and the banking unions at national level, and agreed in 1970. The scheme involved joint management-union agreement on the grading of jobs. By the 1980s the unions had come to believe that the scheme required substantial revision, primarily because of technological change. However, the employers did not wish to renegotiate the scheme, and so far have refused to do so. The collapse of the Federation while BIT was being implemented leaves open the possibility of the scheme being revised at the level of the individual bank. One clearing bank has already introduced a new job evalution scheme, and it is likely that the others will follow.

A consultative committee, the Joint Standing Committee on New Technology, already existed at the case-study bank before the launch of BIT, with regular meetings between management and representatives of the staff union. Items for discussion were raised either by management or by the unions. The BIT project was discussed to some extent, but as regular meetings were suspended in 1987 because of an overtime ban by employees in connection with a pay dispute, there was little opportunity for discussion of the project at a major stage in its development. The bank also had a joint management-staff union safety committee, which met regularly. Issues relating to the health and safety aspects of increased use of VDUs were raised here. Special management presentations on the BIT project were also made to union representatives. Overall, the major consultations over the BIT project took place within the existing pattern of consultative arrangements at the bank. New technology was recognised on all sides as an enterprise-level issue, so there was no question of industry-wide institutions being involved.

The process of consultation on new technology in the bank was highly centralised, reflecting the bank's *modus operandi*. There were no regional, departmental or branch committees. The management representatives on the committees were drawn from the UKRB central personnel department. The union representatives were full-time officers from the union head office, who reported back to the membership through regional officers and the union newspaper. Lay union representatives were involved in the discussions, but the major roles were played by full-time officers.

The process of consultation between management and unions did not result in any major modification to management plans. Up to the time of writing there has been no dispute connected with the project, although labour relations have been tense with the overtime ban in connection with a pay dispute. (In the 1970s the introduction of ATMs did result in conflict, although not in strike action.) However, the process of installing the equipment and introducing new methods of working is not yet complete. Overall, bank management have succeeded in carrying through the project without serious conflicts with employees or with the unions involved; there has been no formal agreement on the new system, because management do not regard a formal agreement as appropriate. Although the failure to obtain such an agreement represents a rejection of BIFU policy, the majority of bank employees seem willing to accept the management approach: this acceptance appears to be based on a traditional loyalty and a fatalistic acceptance that there is little employees can do about new technology. The union role has been primarily one of careful monitoring, and making the membership aware of the significance of the developments taking place.

3.3 Consequences for the workforce of introducing the new technology

3.3.1 Job security

BIT is expected to save 1,500 jobs in the long run. The new system involves less clerical time in processing cheques and money, finding information on screen, producing corporate costings and other financial data, and on the telephone, with quicker access to information. At present, however, employment levels in the bank are still rising, although much more slowly than in the 1970s; it is likely that they will continue to rise slowly in the immediate future, before falling in the 1990s. The level of activity in the industry has increased, and the range of services widened. The development of credit cards and EFT/POS systems has not prevent a large increase in the volume of cheque transactions: the use of ATMs has involved an increase in the number of transactions, not the automation of the existing number of transactions. Bank management argue that new technology makes it possible to adopt a new concept of banking, marketing financial services instead of looking after cash: back-office automation makes front-office marketing possible, and marketing is a labour-intensive activity.

Management undertook not to resort to compulsory redundancy as a direct result of the introduction of the BIT project. There was no explicit guarantee that compulsory redundancy would never occur, for example as a result of branch reorganisation, but there was no expectation that it would be required. The bank continues to experience relatively high levels of labour turnover, so can achieve any require reduction in employment through varying levels of recruitment, voluntary early retirement and, in exceptional circumstances, voluntary redundancy.

Reorganisation in the data processing area has raised the issue of voluntary redundancy for a small group of data processing staff, but not in connection with the BIT project itself. In the long term, the major threat to staff levels could come from the development of EFT/POS, and subsequent contraction in the volume of cheque clearing.

3.3.2 Work organisation and working conditions

According to the staff union, BIT will lead to a bifurcated labour force, composed of "salespeople and terminal tappers". Management accept that there will be changes in the structure of the labour force, but not the perjorative overtones of the union's analysis: for selling read marketing, for terminal tapping read data input. Technological change has not brought about greater union or worker participation in the design of production systems: there was no substantial involvement before or after the launch of BIT, although extensive discussions have been held amongst union members.

Changes in the bank's overall strategy, including the increased used of computers and the reduced use of paper, have resulted in a major change in the physical arrangements of bank branches, with reduced space requirements for cheque and cash processing, record-keeping, etc., and for storage, and an increase in the area used for customer service. This has resulted, according to management, in a better working environment. Management have sought expert advice on lighting and on the ergonomic problems involved in VDU operation. The staff union has been particularly concerned about the possible impact of increased use of VDUs on working conditions, including the possible effects of radiation on pregnant women. The pre-existing consultative committee and safety committee have continued to monitor the problems involved, the BIT project not requiring special arrangements. During the period of developing the BIT project the bank's *Health and Safety Manual* was extensively revised, following discussion in the safety committee. The revision was not directly linked to the development of the BIT project, but dealt with many issues arising from it.

These changes have been tacitly accepted by workers' representatives, without detailed agreement: the bank employees have basically agreed with management's view of the need to introduce new technology and therefore with the accompanying changes in working conditions and working practices.

Both the staff union and BIFU are worried that bank management might further reduce the number and significance of bank branches: since 1975 the number of branches has dropped slightly. They link this anxiety with the bank's earlier purchase of a number of agencies, which the unions think the bank will develop into ancillary bank branches: such agencies have very different working practices and conditions of service from banking, including no union recognition. If agencies were to be linked into the computer system via the BIT project, there would be an obvious potential threat to the existing branch structure. Bank management deny any such intention, and with regard to BIT itself have reassured branches that there will be no "concentrator" branches, with the potential to develop into a two-tier system.

3.3.3 Payment systems and income protection

There has been no alteration in payment systems as a result of the BIT project. Most clerical employees are covered by the existing job evaluation scheme, which has not been revised in the light of the introduction of new technology. Trade union pressure for revision of the job evaluation scheme, justified partly by the expansion of new technology, is likely to continue, however. Although the introduction of BIT has implications for the work of supervisory and lower managerial staff, there is no proposal to change the payment system for such staff. The central importance of computer personnel in the operation of the bank has been recognised by a significantly above average annual pay settlement, and by proposed major changes in the pay structure of data processing personnel, which are the subject of separate negotiations.

Workers have not been explicity downgraded as a result of BIT, so the issue of income protection has not arisen, at least in the short term. Changes in the bank's commercial strategy and in branch organisation may have significant effects on promotion prospects in the future, but such changes are not the result of new technology in itself.

3.3.4 *Training and retraining*

Management regard training as a major priority. The training programme has two strands. The first relates to familiarisation with the equipment. This involves a machine instructor being sent to each branch to give instructions to the supervisor and two other members of staff on fault procedures and to provide familiarisation training to as many people as possible in a period of up to four days. It is expected that familiarisation will gradually cease to be a problem as the normal operation of branch transfers leads to employees familiar with the new systems being transferred to new branches.

The second strand relates to data loading and applications. The basic principle of training in data loading is to use the technology itself, rather than to rely upon booklets and instruction manuals. An interactive video was produced to provide the basic concepts of the new system, and the overall approach involved in BIT. Packages were developed for use on the system itself to provide training in the practical skills required.

The unions have not been involved in the formulation of training programmes, selection of trainees or the conduct of training. Management have selected trainees as required.

3.4 *Effects of the new technology on labour relations*

3.4.1 *Effects on the structure of the workforce*

As already mentioned, BIT is expected to save 1,500 jobs in the long run. More immediately, it will have a significant impact on the structure of the labour force, with more bank employees involved in marketing the bank's various financial services, and more involved in data input. The changes will not represent a straightforward polarisation of the workforce, nor, of course, a shift across the blue to white-collar line. However, they will involve a move away from the traditional conception of the bank employee.

Major changes in the career prospects of bank employees are occurring at the present time: traditional expectations of promotion to the appointed grades and thence to management are becoming increasingly difficult to meet. In pursuing a "niche" marketing strategy, directing efforts to particular groups, the bank is moving away from the all-purpose branch bank. Branches providing more restricted services need fewer senior managerial staff. The development of BIT facilitates this trend, by making available at different levels of the organisation data on individual customers: a niche marketing strategy needs access to clients' accounts in total, not to individual accounts only. BIT enables this to be achieved.

The effect of technological change in baking is probably to increase, rather than reduce, mobility between banks, particularly for technical staff. In part this is the result of a shortage of such specialists. More generally, as accounting processes are automated, and the specific procedures of individual banks are incorporated into explicitly defined computer system routines, the need for extended experience within a specific bank will decline. Bank operatives will therefore not be "locked in" to a specific employer – nor will individual banks be so reliant upon their existing staff.

3.4.2 *Effects on the unions*

Bank employees have traditionally accepted the need for technological change and have not perceived it as a threat. The two unions in the case-study bank both accept the need for technological change, but with varying degrees of enthusiasm. The general banking union, BIFU, has a model agreement on new technology, the first clause of which reads: "The Bank and the Union agree that the most effective available systems and equipment should be adopted, with the aim of providing improved job opportunities, higher rewards and a reduction in working hours for employees, within the framework of a more prosperous industry." To achieve this BIFU argues that the introduction of all new systems should be subject to "mutual agreement" between the bank and the union. Negotiations should take place on all relevant matters, including manpower planning; establishments, workloads and working methods; recruitment and selection; training and retraining; career progression and career development; redeployment and changes in terms and conditions of employment. Among the union's objectives are the use of new technology to create new jobs, an "orderly progress" towards a 28-hour, four-day week, with a minimum of five weeks' holiday, and voluntary retirement at the age of 55. At the time of writing union policy is being reconsidered as it is recognised that the model agreement represents an unrealistically high aspiration, which the union is incapable of realising. Future policy is likely to emphasise the role of consultation more, the role of negotiation less.

BIFU believe that technological change in general, and BIT in particular, pose more threat to their members' interests than bank management claim, especially in the long run. They point to the reduction in the number of clerical jobs enshrined in the cost justifications for the BIT project, a saving of 1,500 jobs. Moreover, they favour new technology as a means of coping with the work pressures caused by the failure to recruit sufficient employees before its introduction. Jobs savings thus lead to new technology, not new technology to job savings.

The bank staff union has a more positive evaluation of technological change in general, and of BIT in particular, seeing technological change as necessary to maintain the bank's competitive position, and reassured by the management's promise of no redundancies.

Neither union has gained or lost members as the result of the new technology, nor has there been any change in demarcation between them. New technology might have been expected to draw the two unions together, in joint negotiations or consultations with management, but no joint activities have taken place.

Both unions operate in a relatively centralised way, with full-time head-office officials playing the major role in negotiations and consultations. In this respect union practice reflects bank management practice. Head-office officials have thus played the major role in consultations with management on the new technology. In part this reflects lay members' lack of interest in and lack of knowledge about the overall pattern of technological development in the bank. More specifically, the BIT project affects all branches in UKRB, and it is therefore appropriate that the major responsibility for responding to management initiatives should rest with central officials. In general, technological innovation might be expected to increase the importance of central officials, since lay members are unlikely to have access to the relevant technological know-how.

3.4.3 *Effects on management*

Major changes in management structure are occurring while the BIT project is being implemented. The changes are due to competition and management corporate strategy, not to the new technology itself, but the new technology is an element in the new management strategy. Overall, bank management are attempting to decentralise operational decision-making, while relying upon financial disciplines to ensure corporate control and profitability. Hence the overall reorganisation of the bank into five groupings, with increased responsibility for individual sections within each grouping.

Within UKRB increased responsibilities in personnel matters, for example, are being given to the regions. Such decentralisation is seen as desirable to reduce costs and increase competitiveness. The policy is made more practicable by the development of information technology, which makes it easier to reconcile the conflicting demands of operational decentralisation and continuing central financial control. In this context the BIT project plays a major role in current changes in the structure of management: it is not an independent causal influence.

In general, British research indicates that technological changes do not increase the importance of the personnel function: personnel issues are considered only during the later, implementation stages of technological innovation. However, the case-study bank operates a clearly thought out and sophisticated personnel policy, which influences all aspects of practice, including the introduction of new technology. The bank has a well-thought out management development programme. The central importance of training is recognised, and the transfer of employees is used as a means of disseminating knowledge about new working practices throughout the bank. In the short run the introduction of BIT may increase the power of the personnel function, as personnel problems may be caused by the transition to the new system, but the increase in influence is likely to prove temporary.

3.4.4 *Patterns of negotiation and consultation*

During the period of the case-study labour relations in the case-study bank were tense, due to conflict over the annual pay negotiations and the London allowance. Such conflicts occurred against the background of a move away from the traditional conception of bank employment as a "career for life" towards more market-oriented employment relationships. The introduction of BIT has not been specifically caught up in such conflicts at the time of writing. As already seen, bank employees basically accept the need for new technology. As the staff union publicly stated, "so far the staff have failed to raise one single obstacle to the changes demanded by the bank". The introduction of BIT has therefore not resulted in any increase in conflict. The overtime ban led to the temporary suspension of the Joint Standing Committee on New

Technology, but overall the BIT project has resulted in an increase in the frequency of management-union contacts. Such contacts have remained at the central level, as in previous discussions.

3.4.5 Conflict

As stated, so far the BIT project has not led to any increase in conflict in the bank, but conflict has been increasing for other reasons. Changes in the bank, including the introduction of new technology, might be expected to increase union militancy in the long run, especially the possible restrictions on promotion. As one interviewee stated, the bank can expect problems from the 26-year-old clerk who finds that his or her prospects of promotion are limited because of changes in branch organisation and working practices. However, such dissatisfactions are likely only in the medium and long term. Moreover, personal discontent among bank employees has not traditionally led to collective action: bank employees have not traditionally been highly "unionate", especially in the case-study bank. Translating individual discontent into union consciousness and action would require major effort by the two unions in the bank.

The effect of new technology upon the union's ability to strike is ambiguous. Three groups need to be distinguished: computer personnel, back-office employees, and front-office employees. Reliance upon a central computer increases the bank's dependence upon central computer personnel: industrial action involving central computer personnel would have a direct and immediate impact upon the bank's operations. The central computer personnel are well organised. To counter any threat to the central computer operations, the bank has treated computer personnel generously in recent settlements, including a major pay restructuring.

The effect of new technology on back-office employees – employees involved in money and cheque clearing, routine account preparation, etc. – is to reduce the importance of "idiosyncratic" knowledge: the clerk's knowledge of the files and procedures is replaced by the application of routine procedures and explicit system routines. Such automation reduces the ability of the individual employee or group of employees to exert influence on the bank through strike action: other personnel, including supervisory and managerial staff, are able to carry out the routines involved. At the same time, the reduced number of employees engaged in such operations may make their replacement difficult simply because of pressures of work. (Work intensification may increase worker dissatisfaction; it also increases management dependence.) Moreover, increased reliance upon automated systems may increase management exposure to industrial action short of stike action: banks are very exposed to errors in data input.

The effect of new technology on the ability to strike of front-office employees – employees directly engaged in servicing the public – is likely to be limited. The BIT project involves making comprehensive information on individual customers avilable to front-office personnel on screen, and their work would be impossible without up-to-date data. However, they themselves have little bargaining power, and new technology is unlikely to change this situation.

3.5 Evaluation

The BIT project is still continuing, and is not due for completion until 1989. A full evaluation of the project is therefore impossible at this stage. However, an interim assessment is possible. Five general comments are relevant.

First, new technology cannot be considered in isolation from other changes taking place within the bank. Competitive pressures are changing the industry, and the case-study bank is changing its overall strategy and structure, as well as introducing new technology. The strategy is increasingly on of "niche marketing", concentrating on those areas of activity where the bank believes it has a competitive advantage. The structure is increasingly decentralised, for operational purposes. The BIT project plays a major role in such changes, but it is not cuasing them, nor is it directly caused by them. Consideration of the significance of the BIT project as a whole would require consideration of the niche marketing strategy and a more decentralised system of operations.

Second, the introduction of new technology in the bank was a management initiative, based on management assessment of future branch organisation needs. Neither unions nor workers' representatives had any role in the discussions on the need for new technology, the basic form of the new system, or the equipment purchased. Once the project had been decided upon management discussed the project in the existing Joint Standing Committee on New Technology, and made special presentations to union representatives. The consultative process was centralised within UKRB, in line with the bank's traditional method of operations. There was limited consultation, but not negotiation, over new technology. Neither union felt they had had any significant input into the BIT project.

Third, both management and unions agree that the central issues of concern to bank employees are job security and training. On job security, the management guaranteed that there would be no compulsory redundancies as a direct result of BIT: changes in employment levels could be accommodated by voluntary early retirement and variations in recruitment. Nor has there been any widespread need for voluntary redundancies at the time of writing. The bank has paid close attention to training, establishing a programme for individual training at branch level.

Fourth, both bank unions are unhappy with the procedures adopted by the bank. BIFU in particular wished to negotiate a new technology agreement, which the bank refused to do. The staff union wanted firmer guarantees on job security than the management provided. The staff union also fear that the BIT project is part of a strategy aimed ultimately at transforming the bank into a financial services agency operating a flexible non-union shop. However, such anxieties are not widely shared by bank employees, even bank employees within the unions. Bank employees have not placed any obstacles in the way of the BIT project; basically accepting management strategy, and the guarantee of no compulsory redundancies.

Finally, the case-study bank has a well-considered personnel strategy, and consider the personel issues raised by new technology within the context of that strategy. The size and financial resources of the bank make it possible to develop personnel strategies on a long-term basis. A member of the project team was specifically charged with responsibility for personnel issues. The major personnel issues considered were job security, training and the heavy work load likely to be required during the transition period. So far the issues have been dealt with successfully, although the peak workload has not been reached. The introduction of BIT was not disrupted by the general labour relations problems the bank experienced in 1987, with the overtime ban in connection with the annual pay round.

Overall, management are introducing the BIT project according to their initial intentions: labour relations issues have not raised major obstacles. The role of the unions in the introduction of the new system has been limited, although bank management have consulted with the staff union involved. This has been due to the relative acquiescence of bank employees in major technological change. At one level, this represents a satisfactory outcome: if bank employees acquiesce in technological change it is inappropriate for the unions to oppose it strongly. However, the importance of BIT lies in the long term, with potential major significance for the operation of branch banking: in the long term passive acceptance may prove an insecure basis for operations, especially in a sector experiencing major competitive pressures and increasingly dependent upon the successful use of advanced computerised systems.

Note

[1] This report is a joint effort, Michael Noon being responsible for carrying out the printing case-study. I am very grateful to the managers and union officials in the case-study firms for their cooperation in the research project: to preserve the anonymity of the research sites it is only possible to thank them generally rather than by name.

7

Technological change and labour relations in the United States

*Daniel B. Cornfield**

1. Case-study of newspaper composing rooms in Chicago**

The transition from hot-metal to cold-type typesetting has been accompanied by a shift in the control of work away from workers and towards management in newspaper composing rooms in the United States since the Second World War. The supplanting of the linotype machine by computerised photocomposition has all but eliminated specialised composing room skills, substituting general keyboard skills for them and generating absolute declines in employment and union membership of newspaper typographical workers. Once the exclusive province of typographical workers, the newspaper composing room is increasingly bypassed altogether by non-union editorial and managerial personnel. Ironically, typographical workers, at least the unionised workers in major metropolitan areas, have succeeded in strengthening the contractual measures for preserving their job security, and have minimised technological displacement of workers throughout the transition to cold type in return for accepting technological changes.

The purpose of this case-study is to explain how job security measures were strengthened during the transition to cold type. After discussing general employment and labour relations developments in American newspaper composing rooms, we turn to an analysis of changes in the collective bargaining agreement between the typographical union and the newspaper publishers' association in Chicago. The Chicago case-study emphasises how the two parties facilitated technological change while preventing redundancies through collective bargaining.

* Vanderbilt University, Nashville, Tennessee

** I am gratefully indebted to three official representatives of Chicago Typographical Union No. 16 for agreeing to be interviewed and for providing me with materials on their union and to Shannon Mann for typing the manuscript.

1.1 The context

1.1.1 Technological and labour relations developments in newspaper composing rooms

As the newspaper printing industry in the United States became more concentrated economically, new composing room technologies diffused throughout the industry and composing room employment declined. Through mergers and acquisitions, the number of newspapers per metropolitan area has declined, while group ownership of newspapers by national diversified communications corporations has increasingly supplanted individual or independent single-newspaper ownership. Between 1910 and 1984, the percentage of daily newspapers owned by corporate chains increased from 3 to 66 per cent.[1] By 1982, roughly three-quarters of dailies had switched to offset printing.[2]

Composing room employment declined with the transition to cold type. According to the US Bureau of the Census, the percentage of newspaper employees who were typesetters and compositors declined from 12.6 per cent in 1970 to 4.1 per cent by 1980 (latest year of available data). Between 1970 and 1980, as total newspaper employment increased from 422,657 to 510,125, the number of typesetters and compositors declined from 50,760 to 21,128.[3]

Several studies have suggested that the shift to cold type has deskilled composing room work. The emergence of computerised photocomposition transformed newspaper composition into a "keyboarding" occupation that can be performed by anyone with average typing skills.[4] Composing room tasks that require the exercise of considerable discretion under the linotype technology have now been automated.[5]

However, Wallace and Kalleberg's time-series analysis of newspaper printing occupational wage ratios suggests that most of the deskilling in composing room occupations occurred before the emergence of cold-type technology. The wage ratio is the ratio between the wages of newspaper compositors and the wages of all unionised printing workers. As such, it is used as a proxy for the relative skill level of newspaper compositors. The newspaper compositor wage ratio remained stable between 1931 and 1950, had declined by approximately 10 per cent by 1960, and stabilised after 1960 when computerised photocomposition spread throughout the industry.[6] This suggests that the transition from hand composition to the linotype technology may have deskilled composing room work more than the transition from linotype to computerised photocomposition – a conclusion that is borne out by data on 29 composing room occupations from the *Dictionary of Occupational Titles*.[7] It also seems that post-hand composition technologies have tended to stratify composing room occupations, while the hand composition occupations are relatively homogeneous with respect to the skill levels required.

As the advent of cold type transformed newspaper composing room work into a "keyboard" occupation, women came to account for a majority of American newspaper compositors. According to the US Bureau of the Census, the percentage of newspaper composing room workers who were women jumped from 14.6 to 62.6 per cent between 1970 and 1980.[8] Roos attributes the feminisation of newspaper composing rooms to the changing gender type of the occupation. Typing, she argues, is traditionally considered "women's work", and women entered the composing room as typing skills became more important in typesetting. Moreover, the capacity of the International Typographica Union (ITU) to control access to jobs had declined with the advent of cold type, especially because the union-controlled, six-year apprenticeship programme was less necessary for providing trained composing room workers. Furthermore, as publishers deployed cold-type technology in part to reduce union power, claims Roos, they often replaced male workers with lower-paid women workers.[9]

The case of the *Chicago Tribune* strike in 1985-86, described in greater detail below, illustrates how employers have used gender divisions to weaken the typographical union. Between 1965 and 1975, with the diffusion of cold-type technology, the percentage of Chicago newspaper typographical union members who were women increased slightly from approximately 5 to 7 per cent. These percentages are lower than the national ones because the predominantly male, unionised composing room workforce in Chicago had effectively operated a closed shop and had won employment guarantees and employer-provided retraining on the new technology, protecting their jobs and obviating the need to hire new workers. During the strike of 1985-86, the newspaper replaced many of the strikers with women who were paid roughly half the wage of a union journeyman. The percentage of women in the composing room increased to roughly 30 per cent, as some of the replacement workers remained employed at reduced wages after the strike, and only 60 per cent of the *Tribune* composing room workers were union members.[10]

Notwithstanding the shift in workplace control towards management, much of the technological displacement of composing room workers occurred through attrition, rather than involuntary lay-offs, and many composing room workers were retrained to use the new technology. Detouzos and Quinn have estimated that almost 90 per cent of the newspaper composing room workers who were displaced by the advent of cold type were retrained and remained employed, left of their own accord, or were given monetary incentives to terminate their employment voluntarily.[11]

That few newspaper composing room workers have been involuntarily made redundant is partly attributable to the philosophy and practices of the ITU, the oldest national labour union in the United States. Founded in 1852, the ITU has attempted to retain jurisdiction over new composing room technologies in order to maintain the job security of its membership. Throughout the twentieth century, the ITU has retained this jurisdiction and has effectively controlled the supply of typographical workers to publish through its apprenticeship programme and by gaining stringent union security arrangements with employers.[12]

During the post-Second World War transition to cold type, the ITU accepted technological change in return for job security. Kelber and Schlesinger refer to this stance as the "policy of controlled automation". It consisted of allowing publishers to introduce new technology only after receiving union approval – a measure few American unions have achieved; sharing in the benefits that resulted from the higher productivity of new technology; participating in retraining, by establishing its own training school in 1956; and pursuing no trade-off between economic demands and job security.[13]

Published in 1967, Kelber and Schlesinger's depiction of ITU policy on technological change antedates the unprecedented job security measures the ITU gained in the mid-1970s in many collective bargaining agreements. Consistent with its policy of controlled automation, the ITU won lifetime employment guarantees, which, again, few American unions have gained through collective bargaining, and permitted employers to offer termination incentives, or "buyouts", to workers who wanted to terminate their employment voluntarily.[14]

Between 1971 and 1983, ITU membership declined from 87,000 to 43,000.[15] This decline, and the subsequent merger of the ITU with the Communications Workers of America (CWA) in 1987, have contributed to what Wallace has identified as a pattern of "quasi-craft unionism" and a trend toward industrial unionism in the newspaper industry.[16] During the post-Second World War era, the number of newspaper craft unions declined as they merged with larger communications unions, reflecting the corresponding increase in the concentration of capital ownership and in employer power in the newspaper industry.[17]

Through its policy of controlled automation, then, the ITU protected its membership from involuntary redundancy as newspaper composing room employment declined and publishers made the transition to cold type. During the post-Second World War transition to cold type in Chicago newspapers, Chicago Typographical Union No. 16 achieved an increasingly complex entitlement to employment through collective bargaining. The continuous strengthening of this entitlement, which occurred through the addition of increasingly stringent measures to the battery of existing job security measures, shows in detail how the ITU policy of controlled automation was locally implemented. The source of data, other than the collective bargaining agreements, is 11 personal and telephone interviews which I conducted with three official representatives of Chicago Typographical Union No. 16.

1.1.2 Labour relations in newspaper composing rooms in Chicago

Chicago Typographical Union No. 16, or Local 16, was founded in 1852 and affiliated to the ITU in 1875. As a craft union, Local 16 membership has comprised only the typographical workers employed in the commercial printing shops and newspapers in the Chicago metroplitan area. Members of other printing crafts are represented by other unions. Local 16 became a CWA local union with the ITU-CWA merger in 1987.

Local 16 had engaged in collective bargaining for newspaper typographical workers with the Chicago Newspaper Publishers' Association or with predecessor employers' associations since 1892. After the Second World War, Local 16 renegotiated master contracts with the Association; these covered all newspaper typographical workers and were achieved independently of collective bargaining between the publishers and other newspaper craft unions. Collective bargaining between Local 16 and the Association ended in 1985 when the Association was disbanded during an impasse in labour negotiations.

Between 1945 and 1985, the number of publishers affiliated to the Association declined from six to two, as several newspapers were purchased by other publishers and other, more specialised newspapers left the Association, which had inadequately addressed their unique needs in labour negotiations. Much of this decline occurred before 1960. After 1960, the Association mainly consisted of two publishers, Field

Enterprises and the Tribune Company. The total number of daily metropolitan newspapers published by them has fluctuated between two and four since 1960.

Local 16 conducted only one strike against the Association between 1945 and 1985. The 22-month strike, which ended in September 1949, concerned the closed shop provision in the contract between Local 16 and the Association. This provision made newspaper composing room employment conditional on membership in Local 16. In 1947, the Taft-Hartley Act prohibited closed shop agreements. The Act was opposed vehemently by organised labour; the ITU refused to relinquish its century-old tradition of maintaining closed shop agreements and was one of a few national unions that attempted to defy the Act. In Chicago, negotiations over the Local 16-Association contract reached an impasse on the issue of maintaining a closed shop. In November, Local 16 members voted 2,330 to 61 in favour of striking and approximately 1,500 typographical workers walked off the job.[18] The 1949 contract that settled the strike omitted the closed shop provision but included a new provision that recognised the union as "the exclusive bargaining representative of all employees covered by" the agreement. The 1949 contract also carried a new provision that effectively continued Local 16 control of labour supply by requiring the union to "endeavour to furnish ... as many [union] members ... as are called for by [a publisher] to perform the duties which are enumerated ... in this Agreement."[19] In 1973, a union shop provision was added to the contract.

Between 1949 and 1973, Chicago newspaper composing rooms none the less operated effectively as closed shops, with all the workers remaining members of Local 16. Complete unionisation of the composing room workforce resulted partly from the traditionally strong pro-union sentiment among these workers, partly from the contractual requirement that the foreman be a Local 16 member (this is permitted under the Taft-Hartley Act), and partly from the direct encouragement of new employees to join the union by the publishers, who wanted to avoid any skirmishes with the union and often escorted new employees to the union hall to ensure they joined the union. Since 1949, the only strike in which Local 16 particpated was a seven-month strike at the Chicago Tribune in 1985-86 (see below).

1.2 The process

1.2.1 Decision-making regarding the introduction of the new technology

During the post-Second World War era, Chicago newspapers introduced four new composing room technologies, constituting a shift from hot-metal to cold-type typesetting. First, teletypesetters were introduced in the early 1950s. Second, photocomposition of display advertisements, a composing room task that has continued to employ approximately two-thirds of the composing room workforce, was introduced in the early 1960s. Third, computerised typesetting of news matter, on which about one-third of composing room workers have been employed, was introduced in 1968. Finally, video display terminals, which are used for entering news and advertising matter into the computer for photocomposition, were introduced in 1975. Each of these technological changes was implemented almost simultaneously by the Association-affiliated newspaper publishers.

For each technological innovation that occurred in the Chicago newspapers, the publishers determined unilaterally and without union consultation which type of technology was to be deployed. Ever since the Second World War, Local 16 has retained veto power over the introduction of new technology and, as discussed below, has focused its efforts on securing jurisdiction over the work performed with new technologies and maintaining job security for its membership. In securing jurisdiction over new technologies, Local 16 has had little or no conflict with the other Chicago newspaper craft unions in determining jurisdictional boundaries.

1.2.2 Negotiations over control of the workplace

Beginning with the introduction of teletypesetters in the early 1950s, Local 16 has attempted to extend its exclusive jurisdiction over the new composing room work that accompanied technological change. Control of composing room jurisdiction was achieved in three ways. First, Local 16 controlled the introduction of new technology through the "substitute processes" provision in its contract with the Association. This provision, first negotiated in 1949, prohibited the publishers from introducing new technologies during the life of the contract; guaranteed that such prohibition did not constitute a waiver of union jurisdiction over any new technologies; and permitted publishers to introduce new technology during a strike or other interference with production.

Modifications of the substitute processes provision increased managerial flexibility in the timing of technological change. By 1958, when negotiations centred on the determination of union jurisdiction over display advertisement photocomposition, the Association had gained the right of mandatory immediate negotiation with Local 16 over the introduction of new technology *during the life of the contract*. However, this new provision prohibited the deployment of new technology during the life of the contract if such negotiation resulted in an impasse.

In 1971, after the introduction of computerised typesetting of news matter, both parties gained the right of automatic appeal to the Joint Standing Committee in the event of an impasse in bargaining over the introduction of new technology during the life of the contract. The Joint Standing Committee, consisting of three Local 16 representatives and three Association representatives, made the final decision by a majority vote. If the Committee failed to agree, a mutually agreed upon disinterested person or an arbitrator from the American Arbitration Association would be appointed committee chair. In 1971, publishers also lost the right of unilateral deployment of new technologies during a strike or other interference with production. Neither Local 16 nor the Association has utilised the appeal procedure.

The second way in which Local 16 attempted to control changes in composing room work was by receiving advance notice of publisher intentions to introduce new technology prior to negotiations on the determination of union jurisdiction. At first advance notice provisions were added to the contract for specific categories of new technology, but in 1971 it was stipulated that 90 days' advance notice should be given of any new composing room technology.

A third mechanism was to extend the jurisdiction to its maximum possible breadth prior to the deployment of the new technology. This was effectively assured by the substitute processes and advance notice contract provisions. Once again Local 16 gained jurisdiction over the new technologies one by one, with a guarantee of jurisdiction over any new composing room technology in 1971.

However, Local 16 achieved less than what it considered to be full jurisdiction in the 1968 computer negotiations, when the publishers won the option of availing themselves of the standard maintenance services provided by the computer manufacturers as part of the leasing or purchase agreement. Prior to 1968, maintenance of all composing room equipment was performed by machinists who were Local 16 members. The publishers argued that the current workforce lacked the requisite skills for computer maintenance. In return for agreeing not to lay off composing room workers because of computerisation during the life of the contract, the publishers won the option of using the manufacturers' computer maintenance services. In practice, computer maintenance has been performed both by Local 16 machinists who have been retrained by the computer manufacturers and by the manufacturers' service representatives.

The 1975 contract, which gave the existing workforce a lifetime employment guarantee (see below), also shifted control over work and labour allocation away from the union and toward the publishers. Under the 1975 agreement, the composing room foreman was no longer required to be a Local 16 member. The publishers had wanted this requirement eliminated in order to increase foreman loyalty to the publisher. Also, transfers of regular situation holders between job classifications, which had been prohibited under previous contracts if extra employees were available to fill the openings in a given job classification, were now allowed, regardless of the availability of extras.

The *Chicago Tribune* attempted to further the trend towards employer control of the workplace during the negotiations over the 1983 contract. Negotiations began in the autumn of 1982 and reached an impasse. The Association, consisting of the *Sun-Times* and the *Tribune*, was seeking the right to make mandatory transfers of composing room workers out of the composing room into other departments. The union resisted this, fearing it would undermine its jurisdiction over the composing room. The Association also sought to eliminate the right of journeymen to employ substitutes, a time-honoured ITU tradition. Furthermore, it was unclear to Local 16 whether or not the Association wanted the make-up of display advertisements to be included in the Local 16 jurisdiction over any new pagination systems in the composing room. Under electronic pagination, which has yet to be introduced in Chicago newspaper composing rooms, display advertisement make-up would be the chief remaining humanly performed composing room task. Of the two Association members, the *Tribune* was more intent on making these demands. It appeared to Local 16 that the *Tribune* was attempting to break the union. The *Sun-Times*, in contrast, was willing to settle with Local 16 and wanted to avoid a strike. One reason was that, unlike the *Tribune*, the *Sun-Times* was making no effort to modernise its production facilities. The *Tribune* had begun construction of new printing facilities by 1984 and had brought in a new, Southern, anti-union management in about 1982.

In July 1985, the impasse persisting, Local 16, along with Web Pressmen's Local 7 and Mailers' Local 2, whose jurisdiction would also be threatened by the institution of the mandatory transfers measure, went on strike at the *Tribune*. This was the first strike in which Local 16 had participated since 1949, and the first joint action of any kind among Chicago newspaper craft unions since the Second World War. Some 1,000 workers from the three craft unions walked off the job at the *Tribune*. The *Tribune* hired strike replacement workers, many of whom were women and were paid wages at roughly half the union rate. The strike, however, was less effective than the unions expected because the Teamsters Union, which represented the newspaper delivery truck drivers, did not honour the picket line.

Local 16 workers returned to work at the *Tribune* with no contract in February 1986. Under National Labor Relations Board (NLRB) law, strikers are ineligible to vote in union decertification elections within one year of striking (a decertification election organised by the NLRB, at the request of the employer, with a view to determining whether a majority of workers still want the union to be recognised by the management for collective bargaining purposes). Anticipating the possibility of a decertification election in which only the *Tribune* strike replacement workforce would be eligible to vote, Local 16 made an unconditional offer to return to work, an offer the *Tribune* was compelled to accept lest it be charged by the NLRB with committing an unfair labour practice. Returning to work with no contract, the Local 16 workers have continued to comprise only 60 per cent of the *Tribune* composing room workforce, the remainder being the non-union, strike replacement workers. The Pressmen made an unconditional offer to return to work one or two months before the Local 16 offer, and several Pressmen members have returned to the *Tribune*. The *Tribune* rejected the Mailers' unconditional offer in 1987 to return to work. At the time of writing contract negotiations between Local 16 and the *Tribune* remain stalled.

Local 16 negotiated a separate agreement with the *Sun-Times* in October 1985, by which time Local 16 had gained the right to bargain with individual publishers and the Association had been dissolved. This contract included the 1975 Supplemental Agreement with the lifetime employment guarantee and incorporated none of the demands that had been made by the *Tribune*. Journeymen retained their traditional right to hire substitutes and the *Sun-Times* received no right to make mandatory transfers out of the composing room.

The *Sun-Times* contract also contained a supplemental agreement on electronic pagination, which would be applicable in the event of the *Sun-Times* introducing this technology. This agreement defined the Local 16 jurisdiction in pagination operations as including the acceptance and processing of all editorial matter, "regardless of how or by whom prepared and from whatever source", and "keyboarding for entering into the system and all make-up of the individual classified display and display ads". Employees other than Local 16 members were permitted to utilise the electronic devices for several functions, including creating, retrieving and editing matter, electronic merging of finished advertisements into full or partial pages, and entering classified matter other than display matter.

To sum up, employment guarantees for the existing workforce were strengthened as the range of humanly performed composing room tasks narrowed. The strengthening of employment guarantees was, however, accompanied by a narrowing of Local 16 jurisdiction over composing room tasks, a decline in union membership, and a shift in workplace control in favour of the publishers. A resolution to the ongoing impasse between Local 16 and the *Tribune* may determine whether this trade-off between job security and workplace control will continue. The two parties differ in their interpretation of the 1975 Supplemental Agreement, which suspended the lifetime employment guarantee during a strike or lockout "for the duration of such actions". Local 16 contends that the lifetime employment guarantee should have been resumed when Local 16 workers terminated the strike and returned to work; the *Tribune*, on the other hand, maintains that lifetime employment guarantee becomes effective only upon the signing of a new contract. The dispute is now before the US District Court awaiting a decision.

Local 16 collective bargaining on technological change and other issues has tended to occur independently of the ITU. In this respect, Local 16 is similar to other large, metropolitan ITU local unions. Although ITU by-laws have traditionally required ITU approval of a contract before a local union presents it to the employer, the chief way in which the ITU has influenced Local 16 bargaining on technological change is by assisting the local union in the creation of contract language pertaining to the definition and extension of jurisdiction over new technologies.

1.3 Consequences for the workforce of introducing the new technology

1.3.1 Job security

Local 16 has achieved several contractual measures for enhancing the job security of its members. These measures include provision for retraining and training, beginning in 1949; work-sharing, which began in 1960; and employment guarantees, coupled with workforce reduction by attrition, starting in 1968. In addition, employees who terminate employment for any reason have continued to be entitled to their accumulated annual vacation pay on separation.

This progressive accumulation of job security measures first encouraged and then compelled publishers to retain the existing workforce of journeymen and apprentices on their payrolls. While retraining and training measures enabled the existing workforce to requalify for employment as new technologies entered the composing room, work-sharing induced full employment by generating temporary vacancies in composing room jobs. Most recently, the addition of employment guarantees has made employment of the existing workforce a requirement.

The contractual entitlement to employment at a newspaper was strengthened during the transition from hot metal to cold type. Beginning in 1952 with the introduction of the teletypesetter, employees under a Local 16 contract have been guaranteed employment and pay for a full shift. In 1968, immediately before computerised teletypesetting entered Chicago newspaper composing rooms, Local 16 gained an employment guarantee against technological displacement for the existing workforce that was to last for the life of the three-year contract. This contractual provision, which the union gained partly in exchange for relinquishing jurisdiction over computer maintenance, as discussed above, guaranteed that "no regular situation holder ... will lose regular employment as a direct result of the introduction of a computer in the future ... This guarantee is understood to be only for the life of this agreement." Replacement employees, who might be hired in the event of turnover in the current workforce, were not covered by the guarantee.

The 1968 employment guarantee was developed locally without the influence of the ITU and other ITU local unions. In fact, Local 16 had been given an oral assurance of guaranteed employment by a trusted publisher representative during the contract negotiations. None the less, fearing that this representative could be succeeded by someone who would not stand by the guarantee, Local 16 unilaterally entered the oral assurance as the employment guarantee into the contract, much to the chagrin of the publisher representative. This employment guarantee was renegotiated into the next two contracts and lasted until 1975.

The 1975 negotiations centred on the introduction of video display terminals (VDTs) for the entry of news matter by writers and editors directly into the computer, bypassing the composing room. Drawing on recently negotiated contracts in New York City, Seattle, San Francisco and Washington, DC, Local 16 gained a lifetime exployment guarantee for the existing workforce in return for the introduction of the VDTs, which occurred immediately after the contract came into effect, and for relinquishing to the publishers the "initial keystroke" for entering news matter into the computer. The agreement also permitted non-union workers to enter classified advertisement matter, which was received from the customer, directly into the computer, thus bypassing the composing room.

The 1975 "Supplemental Agreement", as it was called, also included mandatory retirement for employees aged 65 and older; employer-funded, lifetime severance and welfare payments for retirees; an indefinite hiring freeze; and the right of publishers to offer "termination incentives", which typically amounted to a worker's annual salary, to employees under the age of 65 who wanted to terminate their employment voluntarily. Consequently, Local 16 newspaper membership declined dramatically after 1975, as discussed above.

Work-sharing, or reduced hours of work, one of the oldest methods used by the American labour movement to prevent technological unemployment,[20] was implemented in the 1960 contract. Job security is enhanced by work-sharing because it allows work to be spread among more workers.

Work-sharing was implemented in three ways in Chicago newspaper composing rooms. First, beginning in 1960, an additional week of vacation was progressively extended to a greater proportion of the composing room workforce. Prior to 1960, all employees were entitled to a maximum of three weeks' annual vacation. Second, the length of the day and night shifts was reduced from seven and a quarter hours to seven in 1968. Third, the number of paid holidays per year was increased from six to seven in 1969 and to eight in 1982.

The degree of work-sharing, then, varied with the prevalence and productivity of new technology. It accelerated during the late 1960s and early 1970s when highly productive and automated technology was rapidly and extensively introduced in composing rooms.

In general, job security measures were successively strengthened in anticipation of the introduction of each new form of technology and the elimination of specific composing room tasks associated with that technology. This progressive strengthening of job security measures prevented technological displacement of Chicago newspaper composing room workers. Since 1949, no Local 16 newspaper worker has been displaced involuntarily by technological change, an enviable achievement for any union in the American labour movement.

None the less, Local 16 newspaper composing room membership, while continuing to account for 100 per cent of the Chicago newspaper composing room workforce until 1985, has declined since the Second World War. Between the early 1950s and 1975 it declined from about 1,400 to 1,200. It then declined dramatically to approximately 500 between 1975 and 1979, during which time computerisation occurred. Local 16 newspaper membership continued to decline to roughly 200 by 1988.

1.3.2 Work organisation

The range of composing room tasks has narrowed since the Second World War. In the 1960s, the advent of photocomposition for display advertisements eliminated the linotype machine and combined into one machine with one operator the use of different fonts and print sizes, each of which had been used previously on separate linotype machines by different operators. However, photocomposition did create dark-room tasks in the composing room. Computerised, hot-metal typesetting of news matter, which entered Chicago composing rooms in the late 1960s, automated justification and hyphenation.

During the late 1970s, the entry of video display terminals and photocomposition of news matter completely eliminated the typesetting of news matter as well as the dark room and all linotype machines. By 1981, the main tasks of Chicago composing room workers had come to centre on the photocomposition of display advertisements, which were still received on paper. News matter is entered by reporters and editors directly into the computer for photocomposition via a video display terminal.

1.3.3 Payment systems

Wages in Local 16 contracts have changed in response to changing product and labour market conditions and have been unrelated to collective bargaining over technological change. Historically, the ITU has recognised "only two classes of labor in union composing rooms, journeymen and apprentices". All journeymen, regardless of their specific job classification, receive the same wage. Night-shift and lobster (third) shift wages slightly exceed day-shift wages. In Chicago, the wages of first-year apprentices are calculated as a percentage of journeyman wages; they progress to the journeyman scale during the course of the apprenticeship.

Entry-level apprentice wages, as a percentage of journeyman wages, increased between 1949 and 1971 from 35 to 60 per cent of journeyman pay, where it has remained to the present. The increase in apprentice pay was an effort to attract more people to the trade. During the period of the increase, the economy was expanding and, with the newspaper consolidations coupled with increased newspaper advertising, the size of newspaper composing rooms was increasing. Employment in an average Chicago newspaper composing room approximately doubled to 500 workers between the early

1950s and the mid-1960s. Consequently, the maximum permissible number of apprentices increased from eight to 43 between 1949 and 1968 and, with the advent of computerised typesetting in 1968, decreased to the current level of 40 in 1971. The 1975 Supplemental Agreement banned the hiring of new apprentices. The increase in apprentice pay thus constituted an effort to attract new people to the trade during an era of composing room expansion.

Real journeyman wages in Chicago newspaper composing rooms have increased throughout most of the post-Second World War era. They increased gradually between 1950 and 1968, jumped dramatically in 1971, and declined slightly after 1973. According to the union representatives interviewed, wages were not explicitly traded off or raised in any collective bargaining negotiations *over technological change per se*. Rather, these wage trends roughly parallel national trends in typographical wages, which tended to increase during expansionary periods of the business cycle.[21]

1.3.4 Training and retraining

In 1949, when Local 16 gained its first retraining provision for teletypesetting, the policy of retraining the existing workforce was a half-century-old tradition in the ITU. In Chicago during the early 1950s, it never occurred to the publishers that workers other than the composing room workers would operate the new teletypesetters. Beginning in 1949, the principle of providing the first opportunity for retraining to journeymen and senior apprentices was applied consistently in the retraining provisions in the contract.

The retraining provisions in the contract were subsequently modified in order to broaden both the conditions under which publishers were to provide retraining and the types of worker who were eligible for retraining. Workers who opted for the new jobs would be paid according to a training pay scale that started at 60 per cent of journeyman pay and progressed in six months to the journeyman pay scale. Publishers reserved the right to employ and train other workers to fill the required number of teletypesetter jobs if there were not enough competent trainees.

The 1971 contractual amendments, which have remained in effect to the present day, increased the accessibility of retraining opportunities still further, and added a retraining programme to be implemented "in the event [a publisher] determines that there may be a reduction in one or more classifications of work and there is a need for more Journeymen in one or more classifications of work". This retraining programme was established in order to protect linotype operators whose job security was jeopardised by the transition to computerised typesetting. In order to reduce favouritism and increase the availability of training opportunities, a Joint Selection Committee, consisting of two publishers and two union representatives, was set up to select workers for a 60-day retraining programme. Job openings were then posted among the retrained workers. The Joint Standing Committee, mentioned earlier, was to address any disputes arising out of the retraining programme.

Most of the retraining opportunities consisted of employer-provided on-the-job training. Often, the equipment manufacturers provided the publishers with trainers. A small minority of Local 16 newspaper members enrolled in the ITU training centre, typically during their vacations.

Many Local 16 members enrolled in the Local 16 training school. Established in 1969, this was the only training school operated exclusively by an ITU local union. Printing equipment manufacturers, who wanted to market their products, donated equipment to the Local 16 training school and often received feedback on performance from the union. Approximately one-third of Local 16 newspaper members, especially those who were unable to receive employer-provided on-the-job training due to time constraints, enrolled in the training school. Members from the commercial printing houses, which were too small to provide their own on-the-job training, were more likely than the newspaper members to enrol. Indeed, the training school, which received no financial support from the newspaper publishers, was funded by local union membership dues and the commercial printing houses. By the mid-1970s, most of the newspaper members had been retrained and their participation in the training school diminished still further. The Local 16 training school ceased offering regularly scheduled classes in 1984, but has continued to offer some classes to the present.

The broadening of retraining opportunities after 1949 was partly a response to the increased prevalence and productivity of new technology. The number of teletypesetters in a typical newspaper composing room was still only six by 1968, because the teletypesetters, which produced six lines of type per minute, were no more productive than a journeyman operating a linotype machine. Immediately after the 1968 contract negotiations, high-speed casting devices and computerised teletypesetting, which automated hyphenation and justification, were introduced, and the number of teletypesetters per newspaper jumped to at least 30. The new teletypesetters were capable of producing 12-15 lines of type per minute. The Local 16 training school was established in 1969 during the rapid expansion of computerised teletypesetting, and the 1971 contractual amendments, which were the first to address the problem of job classification reductions, occurred soon thereafter.

The typographical apprenticeship curriculum also reflected changes in composing room technology, as well as the principle of affording training opportunities to the existing workforce. Traditionally, Local 16 apprentices have been "given an opportunity to work in every department of the composing room". Consequently, the apprenticeship curriculum was modified in order to train apprentices on every new technology that was introduced.

The major changes in the apprenticeship curriculum thus began in 1952 with the introduction of a teletypesetter training option, which was available to those sixth-year apprentices who could type 50 words per minute. In 1960, this option was extended, in the light of the expansion of teletypesetter operations, to fourth and fifth-year apprentices, along with a 65-hour typing course for those apprentices who could type only 35-49 words per minute. Training in hand composition was eliminated from the curriculum in 1954. A typewriter training course became a curricular requirement in 1965, with competent typists automatically beginning teletypesetter training.

In 1971 changes were made in the apprenticeship curriculum that have remained in effect to the present. The apprenticeship was reduced from six years to four because of the reduction in the number of different composing room tasks, and the minimum amount of required training in typing and "keyboarding" was greatly increased, to almost 20 per cent of the total training time. The apprentice continued to spend roughly half of the total apprenticeship in page and advertisement make-up, but "general" training, in such tasks as picking up, sorting, distributing and putting away galleys, was eliminated, and unallocated training time decreased from roughly one-third to one-quarter of the total apprenticeship training time. During this time apprentices had previously been assigned by the foreman to variable tasks and/or opted to learn specialised composing room tasks such as linotype and photocomposition machine operation.

These reductions in general and unallocated training time reflected the publishers' desire to reduce the number of job classifications, to limit what had been the customary practice of journeyman specialisation in a job classification, and, thereby, gain greater flexibility in making job assignments. To the union these curricular changes were consistent with its traditional belief that journeymen should be trained in all facets of composing room work.

1.4 Evaluation

The case of Chicago newspaper typographical labour relations illustrates how the implementation of the ITU policy of controlled automation simultaneously strengthened the contractual entitlement to employment and shifted workplace control towards management. Between 1965 and 1975, as the cold-type technology rapidly diffused into the Chicago newspapers, Chicago Local 16 expanded and strengthened the job security measures in its contract with the Chicago Newspaper Publishers' Association. By 1975, the union had gained a lifetime employment guarantee for the Chicago newspaper composing room workforce. Although the Local 16 collective bargaining stance was developed independently of the ITU, Local 16 developed its stance partly by looking at the contracts achieved by other large metropolitan ITU locals in the United States, which suggests that the Chicago case should be fairly representative of typographical labour relations changes in major metropolitan areas.

Notwithstanding the emergence of a contractual entitlement to employment, Local 16 relinquished some of its control of the workplace to management. Most importantly, the advent of cold-type technology substituted general typing skills for the traditional linotype skills and the specialised knowledge that composing room workers had learned through the union-controlled apprenticeship programme. Increasingly the new technology has allowed management to bypass the composing room in newspaper production.

This exchanging of the entitlement to employment for managerial control of the workplace has been called into question by the breakdown in the relationship between Local 16 and the *Chicago Tribune*. The security of the union itself in the workplace and, therefore, the entitlement to employment, are jeopardised by the efforts of this employer to extend its unilateral control over the workplace. The resolution to this struggle may portend the direction of newspaper craft unionism.

2. Case-study of a die manufacturer

2.1 Technological and labour relations developments in the metalworking machinery manufacturing industry

Both labour and management in the American metalworking machinery manufacturing industry have espoused the modernisation of production technology in order to restore the competitiveness of the industry in the world market. Since the Second World War, the stock of metalworking equipment in the United States had aged in comparison to that of other nations,[22] contributing to a deterioration in the international market position of the American industry and jeopardising industry profits and employment. The acceleration in the diffusion of numerical control (NC) and computerised numerical control (CNC) in metalworking machinery manufacturing, beginning in the late 1970s, is part of the effort to revive the industry. By the early 1980s, the United States had developed a stock of machine tools similar in age to those of other industrialised nations.[23]

Despite the labour-management consensus on modernising production technology, few innovations in labour relations arrangements pertaining to technological change have been achieved in metalworking machinery manufacturing during the 1980s. Labour and management have instead diverged in their ideas for implementing technological change and restoring competitiveness. While organised labour has sought to increase worker control of technological change and worker participation in management decision-making, management has aligned itself with foreign manufacturers, developing co-production, licensing and distribution arrangements, attempted to cut labour costs, and lobbied the Government for protective international trade policies. This divergence, generating plant shutdowns and plant relocations to regions of the United States with lower wages, has led to a decline in employment, union membership losses and, consequently, the jeopardisation of the bargaining power of organised labour. Novel collective bargaining arrangements, then, have been rare in metalworking machinery manufacturing.

In order to analyse the impact of technological change on labour relations in this industry, we begin with a discussion of the diffusion of numerical control in the United States. This is followed by analyses of employment trends, trends in occupational employment, skill and earnings, and recent developments in labour relations.

The metalworking machinery manufacturing industry is both a producer and a user of machine tools. The products of metalworking machinery manufacturing, classified as industry 354 in the Standard Industrial Classification (SIC), include metalcutting machine tools (SIC 3541), metalforming machine tools (SIC 3542), the products of tool and die shops (SIC 3544), and machine-tool accessories (SIC 3545), among others.

2.1.1 The diffusion of numerical control

Numerical control had begun to diffuse into several metalworking industries by the early 1960s. The first few numerically controlled (NC) machines for commercial use had been exhibited at the National Machine Tool Show in 1955. By 1964, 50 firms were selling NC machines.[24] During the next 20 years, NC machines spread slowly among metalworking industries, continuing to account for a small minority of the stock of machine tools in the United States. According to the 13th American Machinist Inventory, the number of NC machine tools doubled to 103,308 between 1978 and 1983. However, the percentage of machine tools that were numerically controlled only increased from 2.0 to 4.7 per cent between 1978 and 1983. Shaiken cautions that these statistics understate the importance of numerical control because the machine-tool base is "artificially inflated by the practice of not junking machine tools after their useful life is over" and because NC machine tools are more productive than conventional ones.[25]

The diffusion of numerical control in the metalworking machinery manufacturing industry slightly exceeded that for all metalworking industries. Between 1976 and 1983, the percentage of NC machine tools in this industry increased from 2.9 to 5.6 per cent.[26] NC metalcutting machines, such as turning and milling machines and machining centres, are the most commonly found NC machine tools in the industry.

The relatively slow diffusion of NC machines is implied by the ageing of the existing stock of machine tools in the United States. Between 1945 and 1978, the percentage of machine tools that were less than ten years old declined from 62 to 31 per cent. This percentage increased substantially for the first time since the Second World War – from 31 to 34 per cent – between 1978 and 1983. In the metalworking machinery manufacturing industry, the percentage of machine tools that were less than ten years old increased from 34 to 37 per cent between 1976 and 1983.[27]

The slow spread of NC machines is also partly attributable to the decentralised structure of the metalworking machinery manufacturing industry. The industry consists of many small, family-owned producers, the large majority with less than 50 employees, who cannot afford the investment in the new production equipment and the retraining of the workforce. NC applications tend to be concentrated in large firms.[28]

Stabilisation of productivity has accompanied the slow diffusion of numerical control and the ageing of machine tools in the metalworking machinery manufacturing industry. According to the US Bureau of Labor Statistics (BLS), the productivity growth rate for the industry averaged 1-2 per cent annually between 1960 and 1978.[29] A separate BLS analysis of machine-tool manufacturing shows that productivity increased through the mid-1960s, stabilised through the early 1980s, and jumped to its highest level between 1984 and 1985. The production worker productivity growth rate in machine-tool manufacturing averaged 1 per cent annually between 1958 and 1985 and 2.2 per cent between 1980 and 1985.[30]

2.1.2 Employment trends in metalworking machinery manufacturing

Since the mid-1960s, analyses of productivity have shown that NC technology reduces machining time. In 1965, the BLS claimed that reductions in unit labour requirements over conventional machine tools varied between 25 and 80 per cent; in

1966, the National Commission on Technology, Automation and Economic Progress estimated that machining time with numerical control was 20-30 per cent of that required with manually operated machine tools; and the BLS, in 1982, stated that machining time with numerical control was 35-50 per cent of that with conventional machine tools.[31]

Notwithstanding the labour-saving effects, analysts of the possible employment effects of numerical control have hesitated to forecast any direct effects of this technology on employment. The chief reason for this hesitancy is that it is virtually impossible to separate the employment effects of technological change from those of other factors, including changes in the business cycle, imports and exports, and the volume of industry production.[32]

Between 1967 and 1986, production worker employment in metalworking machinery manufacturing declined annually, on average, by 0.4 per cent, the same as the national average rate of production worker decline for all manufacturing industries. Rates varied across different industrial sectors within metalworking machinery manufacturing. The rates of production worker decline in machine tools exceeded the national average, while in tool and die shops and accessories production worker employment either increased or declined at a lower rate than the national average.[33]

Technological change is one of a set of factors associated with employment decline in metalworking machinery manufacturing. During the recessions of the early 1970s and the early 1980s, the rate of production worker decline tended to exceed the national average. Given the slow diffusion of numerical control during the early 1970s, it is unlikely that this technological change had a large impact on employment during this recession. However, the recent reversal of the trend relating to the ageing of machine tools, the rapid spread of numerical control and the increase in industry productivity during the early 1980s may have combined with the recessionary economic conditions in this period to reduce production worker employment.

Increasing imports of metalworking machinery to the United States is another factor associated with the employment declines of the early 1980s. Importation of metalworking machinery, especially machine tools, increased during the 1970s and accelerated rapidly in the 1980s. Between 1972 and 1981, the share of US metalcutting machine-tool consumption accounted for by imports increased from 10.5 to 28 per cent; by 1986, imports accounted for 51 per cent. A similar pattern exists for metalforming machine tools. For NC machining centres and turning machines, imports had come to account for 66.1 per cent of American consumption by 1986. On the other hand, American industry has consumed few imported dies, tools and jigs or machine-tool accessories.[34]

The metalworking machinery industries into which imports had made the greatest inroads by 1986 were also those that experienced the greatest rates of production worker decline between 1982 and 1986. Machine tools experienced the greatest penetration of imports into the domestic market and the highest rates of production worker decline. In contrast, tool and die shops and machine-tool accessories had few imports and low rates of production worker decline.

Likewise, the metalworking machinery manufacturing industries that experienced the greatest increases in imports suffered the largest declines in the real value of production during the early 1980s. While the real value of production of all manufactured goods increased annually by 1 per cent on average, between 1981 and 1985, that for metalcutting and metalforming machine tools declined annually by 10.2 and 7.4 per cent, respectively. For tool and die shops and accessories, which experienced relatively few imports, the real value of production either increased slightly or declined slightly during the early 1980s.[35]

The rise in volume of imported metalworking machinery, especially from Japan, has been sufficiently great to motivate the National Machine Tool Builders' Association (NMTBA), the chief trade association in the industry, and other groups to seek government protection from imports. In June 1982, several trade associations formed the Metalworking Fair Trade Coalition in order to research and lobby for international trade policies and laws. In 1983, the Coalition lobbied for an American trade policy towards "dumping" by foreign countries in the United States and the NMTBA petitioned the Reagan Administration, seeking restrictions on imported metalcutting machinery to 17.5 per cent of the value of consumption for a five-year period.

In 1986, the Reagan Administration sought five-year Voluntary Restraint Agreements with the Federal Republic of Germany, Japan, Switzerland and Taiwan, China, covering several kinds of metalcutting machinery, and reached agreements with Japan and Taiwan, China. The Reagan Administration attempted to stimulate the modernisation of domestic manufacturing processes by helping to establish a private National Center for Manufacturing Sciences, which will be partially funded by the US Department of Defense.[36]

Despite the decline of production worker employment in metalworking machinery manufacturing, labour turnover rates in this industry have historically been lower than the national average for all manufacturing industries. Given the high level of skill of the machinists, tool and die makers, and other production workers in this industry, and the overall shortage of skilled workers in the United States,[37] employers have attempted to retain their workforces on the payroll during recessionary periods. Rather than laying off workers, employers in this cyclically sensitive industry have usually adjusted workers' overtime hours to changes in product demand.[38] Between 1967 and 1986, the percentage of weekly hours worked overtime in the metalworking machinery manufacturing industries tended to exceed the national average of approximately 8 per cent; they were also more variable over time than the national average.

Given the employment practices of employers in metalworking machinery manufacturing, business failures and plant shutdowns, rather than lay-offs, are more likely to have been responsible for reducing production worker employment in this industry during the 1980s. Indeed, small businesses, such as those that predominate in metalworking machinery manufacturing, are more likely than large businesses to fold during periods of recession.[39]

During the early 1980s, the rate of decline in the number of enterprises in the import-ridden sectors of metalworking machinery manufacturing was relatively high. Between 1981 and 1985, the numbers of metalcutting and metalforming machine-tool manufacturers declined annually, on average, by 2.4 and 1.2 per cent respectively. In contrast, the number of esterprises in other sectors of the industry, which experienced relatively few imports, increased during the early 1980s.[40]

To summarise, the increased diffusion of numerical control in metalworking machinery manufacturing during the early 1980s may have increased industry productivity and contributed to a decline in the number of production workers. However, the employment decline of this period was more likely the result of business failures and plant shutdowns stemming from recessionary economic conditions and the accelerated increase in imported machine tools to the United States.

2.1.3 *Occupational employment, skill and earnings*

Control of the machine tool, and of work, more generally, has been the central issue associated with the effects of numerical control on metalworking occupational employment, skill and earnings. Who controls the machine tool – a part programmer who is a member of management, or a machinist who might be a union member – has yet to be resolved.[41]

The resolution of this issue of control could affect the employment, skill and earnings of machine-tool operators. According to a 1965 BLS study of numerical control, "many decisions, judgement and setup functions required of a highly skilled, conventional machine-tool operator are shifted to the part programer's job". None the less, the BLS claimed that skilled machinists "are generally reported to be able to acquire competency in these [numerical control] skills readily". Furthermore, the formal educational requirements for NC programming had been lowered by the mid-1960s, partly because computers assisted the part programmer in making mathematical calculations. "Initially," the BLS stated, "a knowledge of college mathematics was thought necessary.

Today, a high school level mathematical background – arithmetic, extraction of square roots, algebra, geometry, and trigonometry is reported sufficient" for NC programming. No survey of which occupational groups perform NC programming in the United States has been conducted. Case-studies suggest that no uniform occupational pattern exists among American firms.[42]

The development of computerised numerical control in the mid-1970s has eroded one obstacle to assigning the programming task to a machinist on the shop-floor. Earlier forms of numerical control could not be programmed at the machine and were typically programmed away from the shop-floor. Computerised numerical control can be programmed at the machine by the machinist. In its 1984 study of six machine shops which utilised computerised numerical control, the US Office of Technology Assessment (OTA) observed that programming by the machinist was "technically feasible", but, in practice, still "in an embryonic form".[43]

None the less, NC tool programmers, the US Census term for part programmers, accounted for a tiny fraction of metalworking machinery manufacturing employment in 1980, the latest year for which data are available. According to the 1970 and 1980 US Censuses of Population, the number of NC tool programmers employed in metalworking machinery manufacturing increased by 50 per cent between 1970 and 1980 – over three times the growth rate of total employment in the industry. By 1980, however, these programmers accounted for only 0.1 per cent of metalworking machinery manufacturing employment. The BLS projects that this percentage will be no more than 0.5 per cent by 1990.[44]

Assessing the impact of numerical control on occupational employment trends in metalworking machinery manufacturing is made difficult by the absence of recent data. The most comprehensive and detailed occupational employment data were collected in the 1970 and 1980 US Censuses of Population and probably antedate any employment effects from the recent increased diffusion of numerical control. In 1984, the OTA stated that it was "too soon to forecast precise numerical changes" in occupational employment resulting from the spread of programmable automation.[45] Attributing employment trends in traditional metalworking occupations – such as tool and die makers and machinists – to technological change during the 1970s is virtually impossible because the type and mix of machine tools, whether conventional or numerically controlled, actually utilised by these occupations is unknown.

Instead, most studies of metalworking occupational employment trends have made employment forecasts rather than analyses of past employment trends. In 1965, for example, the BLS expected the employment of machine-tool operators to decline due to displacement by numerical control, but noted that this decline could be partially offset by the employment of skilled machine-tool operators as part programmers. In 1966, the National Commission on Technology, Automation and Economic Progress projected an employment decline for machine-tool operators due to an expected increase in their productivity from numerical control. In 1982, the BLS predicted that skilled machine operators would continue to be in demand for working with the most advanced NC machines. These forecasts suggest that any employment declines due to numerical control are likely to be greater among less skilled rather than more skilled metalworking occupations.[46]

In contrast, the OTA claimed, in 1984, that "the proportion (and number) of skilled machinists is likely to fall in the long term because of a growing use of NC technology ... This will happen because in some cases NC allows less skilled people to substitute for skilled journeyman machinists in operating and/or programing machine tools."[47]

Most studies of numerical control have concluded that it deskills the machining process and gives management greater control over it. According to Shaiken, with numerical control

> the planning decisions are made away from the shop floor by a part programmer who determines what the machine will do and then translates this information into a form that can be read by the machine. The operating responsibilities remain with the machinist, who is largely reduced to making adjustments if something unexpected happens or stopping the machine if an accident occurs. The worker becomes a monitor rather than a paricipant in the production process.[48]

Indeed, historian David Noble argues that during the late 1940s and early 1950s, the management community embraced numerical control more readily than record-playback, an alternative to numerical control, because numerical control led to greater reductions in the discretion exercised by the skilled machinist and thus in management dependence on the skilled machinist.[49] Furthermore, the BLS and the OTA claim that many firms have acquired numerical control because of the shortage of skilled workers in the United States and the lower skill levels required.

This deskilling of metalworking occupations suggests that occupational wages may be altered by the introduction of numerical control. It is, however, difficult to assess national trends in the impact of numerical control on occupational earnings in metalworking machinery manufacturing for four reasons. First, the extent to which machine-tool operators work with NC, conventional, or both types of machine tools is unknown. Second, the most comprehensive occupational wage data for metalworking machinery manufacturing – that is, BLS data on the wage rates of operators of NC and other machine tools – covers only three metropolitan areas.[50] Third, the most recent BLS data were collected in 1983 and may be insufficiently recent for discerning the effects of numerical control. Fourth, few data purely on metalworking machinery manufacturing are available.

Notwithstanding these limitations, the 1983 BLS wage survey of non-electrical machinery manufacturing is useful for gauging the impact of numerical control on occupational earnings. In Boston, for example, 27 per cent of non-electrical machinery manufacturing production workers worked in shops with no NC machine tools; 9 per cent worked in shops in which the wage rate of operators of NC machines was higher than that of operators of conventional machines; 2 per cent worked in shops where operators of both types of machine tools were paid the same wage; none worked in shops in which the wage rate of NC operators was lower than that for conventional machine-tool operators, and 62 per cent worked in shops that had no formal wage rate policy for operators of NC machine tools. These findings suggest that numerical control has had little impact on occupational earnings. In 16 of the 23 metropolitan areas, at least 61 per cent of production workers were employed in shops with no NC machines or with no formal wage rate policy for operators of NC machines.[51]

The few available detailed occupational wage data for metalworking machinery manufacturing suggest that a polarisation of wage rates among occupations of different skill levels occurred between 1973 and 1983. The BLS collected occupational wage data on tool and die shops and machine-tool accessories in Detroit, Hartford and Los Angeles. In all three cities, the percentage change in the real wage rates of the higher-skill occupations tended to exceed that for the other occupations. Although limited in scope, these findings are not inconsistent with the expectation that numerical control deskills lower-skill jobs more, and therefore lowers the wages for these jobs more than those for higher-skill jobs.[52]

2.1.4 *Developments in labour relations*

As increased imports and international competition have jeopardised profits and job security in American metalworking machinery manufacturing, both labour and management have come to espouse the modernisation of production processes in order to revive the competitiveness of the industry. Notwithstanding their consensus on the importance of technological change, labour and management have made few innovations in labour relations arrangements in order to facilitate technological innovation. Rather, while the International Association of Machinists and Aerospace Workers (IAM), the chief labour union in this industry, is pressing for more labour participation in the design, implementation and control of new production technology, management are attempting to cut labour costs, relocate business establishments in lower-wage regions of the United States, develop joint ventures with foreign manufacturers and import their products to the United States, and research and develop new production technology with partial government support. Managerial initiatives for restoring the competitiveness of metalworking machinery manufacturing have thus emphasised the strengthening of ties between American and foreign manufacturers and between American manufacturers and the American Government. In 1982, for example, the Bendix Corporation, a major machine-tool producer, entered into five co-production and purchasing agreements with Japanese and Italian firms; the Acme-Cleveland Corporation and Mitsubishi Heavy Industries of Japan jointly designed and produced a turning machine, and by 1984 Acme-Cleveland had gained exclusive distribution rights for Mitsubishi machining centres in North America; the Cross and Trecker Corporation began importing Japanese machining centres in 1984. According to the *Wall Street Journal*, machine-tool producers "to an unusual degree ... buy, sell and share technology with scant attention to national boundaries".[53]

The establishment of the National Center for Manufacturing Sciences, initiated by the Reagan Administration in 1986 after domestic producers lobbied for restrictions on imports, illustrates the forging of ties between Government and manufacturers. With initial funding from the US Department of Defense, member firms and the State of Michigan, the Center will promote research and development of manufacturing production processes. As of 1987, Center membership consisted of 70 firms and was expected to increase to 500-800.[54]

As far as their relationship with the labour movement is concerned, management have attempted to wrest wage and benefit concessions from the workers and have begun to relocate out of the north-east into lower-wage regions, especially in the Sunbelt. For example, Cross and Trecker moved some of their Milwaukee production to Georgetown, Kentucky, cutting their hourly labour costs by 40 per cent. Acme-Cleveland gained a union contract that reduced benefits and cut the hourly wages of the least skilled workers from $10.50 to $8.25 in 1982; they achieved more wage and benefit reductions in 1984, and relocated some of their operations in Kentucky, North Carolina and other lower-cost locations. Furthermore, the IAM, claimed the *Wall Street Journal* in 1987, had not pressed for wage increases in five years.[55]

The management effort to cut labour costs, especially by relocating in lower-cost regions, may have contributed to declining union membership during the 1980s. At the 1986 IAM National Machine and Manufacturing, Shipbuilding and Ship Repair Conference, which included IAM delegates from the tool and die industry, the report of the organising committee attributed recent IAM membership losses partly to foreign and domestic relocations, as well as to sell-outs, closures, lay-offs, bankruptcies, employer efforts to eliminate unions, foreign trade and subcontracting. The report also cited several obstacles to union organising, including employer resistance to unionisation.[56]

The few available data on unionisation suggest that the percentage of unionised workers in metalworking machinery manufacturing declined during the 1980s. Between 1974 and 1980 (years of available data), the percentage of unionised workers hovered around 29.5 per cent.[57] Data on IAM membership in the United States and in the tool and die industry, the only available post-1980 unionisation data on metalworking machinery manufacturing, imply that the percentage unionised in this industry declined after 1980. Between 1980 and 1985, IAM American membership, which covers several manufacturing industries, declined annually, on average, by 8.2 per cent; in contrast, manufacturing production worker employment in the United States declined by only 2.3 per cent annually, on average, between 1980 and 1985. In the tool and die industry, production worker employment increased annually by 0.2 per cent between 1980 and 1985, while IAM tool and die membership declined annually by 6.3 per cent during this period.[58]

The IAM policy towards technological change is part of a general industrial policy which is presented in a book, *Let's rebuild America*, published in 1984. Inspired by Western European principles of democratic socialism and a desire to revive heavy industry, achieve full employment and improve the quality of American life, the general industrial policy goals of the IAM are to realise the right of each individual to rewarding employment; an equitable distribution of wealth and political power; industrial democracy in the workplace, including trade unionism and employee participation in enterprise-level and national decision-making; and peace and prosperity. The IAM industrial policy includes a proposal for a national Labor Industrial Sector Board, an independent government agency chaired by a trade unionist who would appoint trade unionists, industrial managers and professionals to 32 vice-chair positions representing different industrial sectors. The Board would be generally responsible for developing, coordinating and directing a national investment and production strategy and achieving full employment. Among the Board's specific tasks would be the stimulation of new technology research and development.[59]

The development of new production technology, according to the IAM industrial policy, would create jobs rather than displace workers. In this regard, the IAM Workers' Technology Bill of Rights includes a call for using new technology in a way that creates jobs; sharing labour productivity gains from new technology with workers; requiring employers to pay a replacement tax on labour-displacing new technology; developing the American industrial base before new technology is exported abroad; retraining and re-employing technologically displaced workers; and giving trade unions the right to participate in managerial decision-making on the introduction of new technology, and workers the right to monitor new technology.[60]

The IAM has also attempted to increase worker control of technological change and prevent technological unemployment through collective bargaining. At the 1981 IAM Electronics and New Technology Conference, the IAM presented model contract language pertaining to technological change and encouraged its local unions to bargain for this language. Some of these model contractual provisions have been used traditionally by the American labour movement,[61] including reducing the workforce by attrition, rather than by lay-off or downgrading, when technological change requires such a reduction; worker retraining by the firm when technological change results in new and/or revised job classifications; seniority-based inter-plant transfer rights, with retention of seniority, for employees who are laid off because of technological change; preventing new jobs created by technological change being classified as non-bargaining unit, and also preventing bargaining unit jobs whose content is altered by technological change being reclassified as non-bargaining unit jobs; and automatic recognition of the IAM as workers' representative in new plants producing similar products to those currently produced in IAM-represented plants.[62] In contrast, the IAM model contract language on advance notice of technological change, and on joint consultation over technological change, are definitely innovative.

Despite its innovative proposals for enhancing worker control of technological change and preventing technological unemployment, the IAM has made few collective bargaining inroads with respect to technological change in metalworking machinery manufacturing, although it has achieved certain traditional contractual arrangements. The few recent available data on IAM contract provisions in this industry were collected in two IAM surveys. The first was an October 1988 survey of all 68 IAM agreements in the tool and die industry. The second covered the six area-wide, multi-employer, IAM master agreements in machinery and manufacturing, which were in effect in August 1986. These six agreements were located in San Francisco, Washington, DC, Seattle, Tacoma, St. Louis and Vancouver. Most of the agreements contained a provision for laying off workers in reverse order of seniority, a procedure traditionally espoused by the American labour movement and commonly found in metalworking machinery manufacturing.[63]

Five of the six master agreements and 97.1 per cent of the tool and die agreements contained this provision. Although severance pay and supplemental unemployment benefits (SUBs) are conventional income security arrangements for displaced unionised workers in the United States,[64] only one of the six IAM master agreements and 16.2 per cent of the tool and die agreements had a provision for severance pay and none had a SUB provision. The paucity of income security arrangements may result from the low lay-off rates in this industry and the employer practice of adjusting overtime hours rather than employment to changes in the volume of production. However, provisions for vacation pay on separation or lay-off were more common. Of the 68 tool and die agreements, 83.8 per cent had provisions for vacation pay on separation and 54.4 per cent had vacation pay on lay-off. Five of the six master agreements had vacation pay on separation and four had vacation pay on lay-off.[65]

The IAM achieved fewer collective bargaining gains with respect to technological change specifically. None of the agreements contained provisions for advance notice of technological change, joint committees, lay-off by attribution, preferential hiring at other company establishments or employer-provided moving expenses. Of the six master agreements, only the Vancouver agreement had a provision for employer-provided retraining, while among the tool and die agreements, only 1.5 per cent had such a provision.[66]

2.1.5 *Summary*

To sum up, as increased foreign competition has led both management and labour in metalworking machinery manufacturing to advocate the modernisation of production processes, technological change has accelerated during the 1980s. Labour and management have, however, diverged as regards the methods for modernising the industry. Management have tended to strengthen their ties with foreign manufacturers, to gain government support for research and development, to wrest wage and benefit concessions from the unions, to relocate in lower-wage regions, and to resist unionisation. In contrast, the IAM has attempted to increase worker control of technological change through national industrial policy and innovative collective bargaining proposals, but has realised few of these proposals. Instead, the management efforts to modernise the industry have led to declines in union membership that may have temporarily eroded the bargaining power of organised labour and prevent innovation in collective bargaining arrangements for adjusting to technological change.

2.2 *Context for the introduction of numerical control at a die manufacturer*[67]

The case of the Durable Dies Company (a pseudonym, as the firm requested anonymity), which introduced numerical control into its production process in 1983-85, is one in which cooperative labour-management relations persisted throughout the process of technological change, notwithstanding the employment reductions and declining proportion of skilled workers that accompanied it. Cooperation between management and workers, who were represented by an IAM local union, endured in part because the job and income security arrangements in the union agreement – arrangements commonly found in IAM contracts in the tool and die industry – minimised the disruption to workers' lives and facilitated worker acceptance of the technological change. Moreover, labour-management cooperation resulted from a traditionally trustful labour-management relationship and the timing of the technological change, which occurred during a post-recession economic recovery when job security and career advancement opportunities were improving at Durable.

None the less, workers feared that the technological change would jeopardise their job security. Such fears may have derived in part from the lack of worker involvement in the making of decisions on the selection and introduction of new technology, which was performed primarily by management. Like most of the IAM contracts at tool and die shops, the Durable contract lacked provisions for worker participation in such decision-making.

2.2.1 *The enterprise*

Typical of firms in the decentralised tool and die industry, Durable Dies is a single-plant, family-owned firm employing about 50 workers, founded in the mid-1930s. It manufactures metal-stamping dies on a custom basis. Its customers tend to be large, corporate, mass production manufacturers of electronic household appliances, who use the dies for stamping out the metal parts of such products as washers and dryers, kitchen ranges and microwave ovens. Prior to the mid-1970s, the firm also produced dies for car manufacturers. However, this side of the business has since declined as car manufacturers are increasingly substituting plastic for metal parts.

Reporting to the president, an older brother in the family that owns the company, are an accounting/personnel department, which consists of two non-union office workers, and a general manager, currently the president's younger brother. The general manager oversees the engineering department which, for the last 20 years, has consisted of four or five non-union die designers (or draughtsmen), who receive the customer's part blueprint and develop the die design for the die makers. Also reporting to the general manager are two shop superintendents, one from each of the die-making and machine sections of the factory where the unionised production workers are employed. The machinists in the machine section of the factory, using part sketches made by die makers, make the die parts, which are assembled into a die by the die makers in the die-making section of the factory.

2.2.2 Labour relations prior to the introduction of the new technology

The production workers at Durable have been represented since 1945 by an amalgamated (that is, multi-employer) IAM local union, which covers all crafts and occupations in the factory and whose membership includes workers in other establishments in the city. Almost all the Durable production workers are union members.

Collective bargaining has typically occurred at the enterprise level, yielding a contract simply for the Durable workers. The union negotiating committee has usually consisted of three production workers from Durable Dies plus a full-time salaried IAM business agent, an employee of the IAM district organisation. Although the national IAM has suggested the desirability of wage standardisation on a national basis, IAM local unions and business agents are typically free to, and do, develop their own enterprise-specific wage scales and working conditions. Collective bargaining at Durable Dies has relied only minimally on input from the national IAM, with the business agent usually providing the negotiating committee with wage and working conditions data from other tool and die shops in the city and data on national trends in collective bargaining on fringe benefits. All contracts must be ratified by a majority vote of the union membership at Durable Dies.

Labour relations have been harmonious under the current family ownership. Since 1965, there have been only two strikes: a six-week one in 1969 and a nine-week one in 1971. The strikes concerned economic issues and occurred at a time when strikes were increasing nationwide in metalworking machinery manufacturing.[68] After 1982, when Durable purchased its first NC machine tool, the number of grievances filed by production workers declined, and no grievance has pertained to numerical control.

2.2.3 General characteristics of employment relationships

Employment practices at Durable are similar to those found generally in US metalworking machinery manufacturing. The firm has rarely laid off workers but has tended to adjust the amount of overtime to changes in the volume of production. According to the union agreement, workers are to be selected for lay-off in reverse order of seniority and, if possible, laid-off employees are to be returned to their former jobs. The agreement makes no reference to technological change in regard to lay-off procedure. The agreement also allows laid-off workers to accept employment in another job classification at the wage rate of that job classification. Income security arrangements at Durable are similar to those found in most of the IAM tool and die shop agreements discussed above, the union agreement providing for vacation pay on separation. The introduction of numerical control was accompanied by no changes in lay-off procedure or income security arrangements.

The system of job classification, compensation and labour allocation partly reflects the production technology. The union agreement designates a separate job classification for the operator of each type of machine tool, with some subdistinctions by operator skill level. In addition, the agreement contains three job classifications for die makers, distinguished by skill level – a separate job classification for "tool try-out" (testing the dies on the presses) – and several non-skilled job classifications. An hourly wage rate is assigned to each job classification, with skilled jobs receiving higher rates than less skilled jobs. An actual production worker wage is largely a function of the job classification, number of regular and overtime hours worked, and the shift on which the worker worked. The number of employees in a given job classification is determined unilaterally by management.

Although the union agreement contains a job posting and bidding system and allows workers to be transferred among jobs with different classifications, movement among jobs, other than the progression from an entry-level job to a journeyman or machinist job classification, rarely occurs. Typically, a worker enters Durable Dies with a high-school diploma as a general trainee, advances to helper and then to die-maker apprentice or machine-tool operator trainee, and finally "tops off" at the full wage rate of the job classification for which he/she was trained. During the four-year die-maker apprenticeship and the two-year machine-tool operator training period, the wage rate progresses from 55 per cent of the full wage rate for the job classification to the full wage rate. Workers who are transferred temporarily experience no pay cut and may receive a temporary wage increase if they are transferred to a higher-wage job classification. Few workers who are being paid at the full wage rate of their job classification opt to change jobs, lest they accept a pay cut by virtue of entering trainee status in order to qualify for employment in another job classification. This system was not altered by the introduction of numerical control.

2.3 The decision-making process

2.3.1 Decision-making regarding the introduction of the new technology

In the autumn of 1982, Durable Dies purchased its first NC machine tool – a travelling wire electrical discharge machine (WEDM) with non-computerised numerical control. Three WEDMs with computerised numerical control were purchased in 1984-85. The decision to purchase and select new production technology was made mainly by the company president in consultation with the general manager and the superintendent of the factory machine section. The president decided to purchase the NC equipment in order to improve product quality and productivity and, therefore, the market competitiveness of the firm. He researched and observed numerical control in operation for six months prior to his first purchase of a WEDM in 1982. Throughout this six-month period, the union was aware that the firm was considering the purchase of new equipment, but it was not involved in the decision to acquire the new equipment nor informed about WEDMs.

2.3.2 The role of negotiation and consultation

Prior to their purchase of the first WEDM, Durable management wanted to ensure that they would be able to programme and operate the new equipment with non-union personnel. Given the large capital outlay involved, and both management and union's lack of familiarity with numerical control, the firm wanted to select the personnel who would programme and operate the WEDM, rather than use the job posting and bidding system in the union agreement. In order to control the programming of the WEDM, the firm opted to purchase a WEDM with non-computerised numerical control rather than one with computerised numerical control, knowing that non-computerised numerical control could not be programmed at the machine on the shop-floor.

In addition, Durable management sought and gained a contractual provision in the union agreement before the purchase of the first WEDM that would allow the firm to programme and operate a WEDM with non-union personnel, at least temporarily, in the event of the firm purchasing a WEDM. There was no contractual precedent at Durable for allowing non-union workers to operate a machine tool or for introducing new production technology. All previous equipment purchases had been of conventional machine tools, with which union and management were already familiar and which were operated only by union machinists.

The union accepted the provision with the informal (that is, oral) understanding that at an unspecified time, assuming the training of workers was feasible and the workload of the WEDM had increased, the WEDM operator job could become part of the union bargaining unit and that the firm would not lay off workers because of the technological change. The union reaction to the new technology was one of ambivalence. The production workers feared that the WEDM would remove work from the other machine tools and jeopardise the job security of the operators of the other machines. However, the union, aware of the presence of numerical control in other die shops in the city, also agreed with management that new production technology could improve the market competitiveness of Durable Dies and thereby enhance worker job security. Furthermore, the union agreed that management should be able to familiarise themselves with the new equipment. The union was unaware of the national IAM model contract language on technological change, including language on advance notice and joint consultation, which had been presented at the national IAM Electronics and New Technology Conference in March 1981, as discussed above.

The WEDM operator job became a job classification in the union bargaining unit about one and a half years after Durable acquired its first WEDM. Initially, the WEDM was programmed by the engineering department and operated in the shop by the superintendent of the machine section. By the summer of 1984, the WEDM workload had become too great for the one superintendent, who was also serving in his supervisory capacity. Union and management agreed to include the job in the union bargaining unit and to continue assigning the WEDM programming to the engineering department. The WEDM operator job was filled by the job posting and bidding system in the union agreement.

Management determined unilaterally the number of employees in the WEDM operator job classification, with the union requesting a "normal" day's workload for the job. Although each of the conventional machine tools had been operated by one worker at a time, the union accepted the possibility that one worker could simultaneously operate more than one WEDM. Since the acquisition of the three CNC WEDMs in 1984-85, the WEDM staffing level has remained at two workers operating four WEDMs during the day shift and one worker operating all four WEDMs during the night shift. The only issue on which union and management differed was the WEDM-operator job wage level (see subsection 2.4.3).

2.4 Consequences for the workforce of introducing the new technology

2.4.1 Job security

The introduction of numerical control at Durable Dies was accompanied by a decline in production worker employment. Assessing the employment impact of the WEDMs at Durable, is, however, complicated by the deline in the car side of Durable's business and the 1982-83 national recession, which immediately preceded introduction of the WEDMs.

The trend in real sales at Durable Dies has tended to parallel that of the American tool and die industry as a whole. The trend in production worker employment at Durable, on the other hand, has deviated from that of the industry. The number of production workers in the tool and die industry declined during the 1982-83 recession but returned to the pre-recession levels in 1984-86, following the national trend in real sales. In contrast, the number of production workers at Durable declined by one-third during the 1982-83 recession, but rose in 1984-87 to no more than roughly 86 per cent of the pre-recession production worker employment level, notwithstanding the return of Durable's real sales to their pre-recession levels. This suggests that the introduction of the WEDMs at Durable in 1983-85 may have led to an increase in worker productivity and, consequently, to a reduction in the production workforce.

The employment reductions at Durable occurred during the 1982-83 recession, before most of the WEDMs were acquired, and with little involuntary displacement of workers. Most of the 17 exits from the bargaining unit occurred through attrition. Of these, four machinists and six employees in the lowest-wage general trainee, helper and service occupation job classifications either resigned from the firm or retired (eight resigned) and two class "A" die makers were promoted into non-bargaining unit jobs. The remaining five (one machinist and four employees in the general trainee, helper or service occupation job classifications) were laid off in the recession and recalled to the firm during the post-recession recovery. Four of these five returned either to their previous jobs or to higher-paid job classifications than previously. One laid-off machinist returned as a die-maker apprentice.

2.4.2 Work organisation

One effect of WEDMs on die-making at Durable was to reduce the task of "hand-fitting", or making blocks that cannot be produced on the machine tools in the machine section, using such manually operated tools as files, profile grinders and air grinders. The WEDM, which can cut more precisely than conventional machine tools, has reduced the amount of hand-fitting by about 75 per cent. With this reduction in hand-fitting, fewer dies are returned for modification after tool try-out: since the introduction of the WEDMs the percentage of cutting dies returned from tool try-out to die maker for modifying has declined from 20 to 5 per cent. This reduction in the amount of hand-fitting has blurred somewhat the skill difference between class "A" and "B" die makers.

Although WEDM programming has remained officially the responsibility of the engineering department and outside the union bargaining unit, the union WEDM operators, while on the job, have informally taught themselves WEDM programming and actually do about half the programming. At the time Durable purchased its first WEDM, the engineering department attended a one-week training session on programming conducted by a manufacturer's representative at Durable. The engineering department continued to do all the programming of the WEDM until the firm purchased computerised numerical control in 1984-85. The three WEDMs with computerised numerical control can be programmed at the machine. Moreover, once an operator sets up the WEDM, the WEDM, unlike conventional machine tools which require the operator to be present all the time the machine is in operation, operates and monitors itself, and shuts itself off if a problem occurs and when it completes its programmed instructions. Being inquisitive and feeling that the WEDM operator job would be more interesting if it were to include programming, WEDM operators acquired a WEDM instruction manual from a manufacturer's service representative and, during the times when the WEDMs were in operation, taught themselves in two or three months how to make simple programmes for the WEDMs.

Management, who were initially hesitant about allowing the WEDM operators to programme the WEDM, agreed with the operators that allocating some of the programming to them would make their job more interesting and improve product quality. Management have since encouraged the operators to learn how to programme the WEDMs. At present, the superintendent of the machine section allocates all WEDM programming tasks to either the engineering department or the WEDM operators, depending on the length and complexity of the programme and the amount of time the engineering department and WEDM operators have available. More recently, and informally, management have begun to teach the WEDM operators how to programme the WEDM with non-computerised numerical control, which cannot be programmed at the machine.

2.4.3 Payment systems and income protection

Initially, management and union disagreed on the WEDM operator wage level, but the two parties were not far apart in their wage proposals. The management developed their proposal from information provided by the equipment manufacturer and by the superintendent who had operated the WEDM. The union researched the wages that were paid to workers in comparable jobs at other die shops in the city. Union and management agreed that the WEDM operator job required less skill than most conventional machine-tool operator jobs. The WEDM operator wage was therefore set at a level below that of all the die-maker and conventional machine-tool operator job classifications and above that of the janitor and the helper job classifications. Other than the establishment of the wage level for the new WEDM job classification, the introduction of numerical control had no effect on the pay system at Durable.

Between 1981 and 1983, nine bargaining unit employees were reassigned to lower job classifications. A majority of the downgraded employees returned to their original jobs during the post-recession recovery. Five were die makers or machinists who were reassigned to lower-wage helper or service occupation job classifications but returned to their original jobs after 1983. The other four were machinists who were reassigned to lower-wage machinist, helper or service occupation job classifications. Of these four, one resigned from the firm, two were promoted into machinist jobs but with lower wage rates than their original job classifications, and one remained in his downgraded status.

2.4.4 Training and retraining

The WEDM operators at Durable Dies were trained to operate the WEDM on the job by the superintendent of the machine section. Originally, the manufacturer's representative who installed the first WEDM taught the superintendent, a former jig bore machinist at the firm, how to operate the WEDM in two days. After the WEDM operator job classification became part of the union bargaining unit, the WEDM operator job was filled from the ranks of union employees through the job posting and bidding procedure in the union agreement. Two of the first three union WEDM operators had followed the usual job progression, moving from general trainee to helper before being promoted into the WEDM operator job classification. The third was promoted directly from general trainee to WEDM operator. All three were trained by the superintendent to operate the WEDM. After their first two or three months of training, the WEDM operators required little or no supervision, and after the first six months they were able to work at an appropriate speed with the desired product quality. The training period for the WEDM, then, was about one-quarter of the time required for training on the conventional machine tools.

2.5 Effects of the new technology on the structure of the workforce

The occupational structure of production employment at Durable shifted in two ways as production employment declined and WEDMs were introduced. First, the occupational structure came to comprise a greater proportion of employment in low-wage job classifications. Second, the ratio of die makers to machinists declined in tandem with the occupational changes by wage level.

Declines in production worker employment in the high- and medium-wage job classifications were especially great during the post-recession recovery. Between 1977 and 1982, with the onset of the recession, the percentage of production workers employed in high-wage jobs increased, while the percentage employed in other jobs declined. After the recession, however, with the introduction of the WEDMs, the percentage of workers employed in the high- and medium-wage jobs declined below the 1977 percentage, while the percentage employed in low-wage jobs increased to a higher level than before. Most of this increase was attributable to the addition of the WEDM operator jobs.

A decline in the employment of class "A" die makers, the highest-wage job classification at Durable, contributed to the decline in the percentage of Durable workers employed in high-wage jobs. Both the decrease in the car manufacturing business and the introduction of the WEDMs may have contributed to this decline. The number of class "A" die makers declined from 13 to 9 between 1977 and 1981, and remained at nine between 1984 and 1987. Dies for car parts are more complex and require greater skill in die-making than dies for electronic household appliance parts. With the decline in Durable's car business in the mid-1970s, Durable thus needed fewer class "A" die makers. As some of them reached retirement age during the late 1970s, Durable did not replace them but continued to employ a total of six to eight class "B" and "C" die makers. The introduction of the WEDMs in 1983-85, moreover, did not generate any increase in the number of class "A" die makers and may have led to a slight decline in the number of class "B" and "C" die makers.

The introduction of WEDMs at Durable also coincided with a shift in employment out of the die-making section of the factory and into the machine section, a trend that had already begun before the introduction of the WEDMs. Between 1977 and 1987, almost 10 per cent of the Durable production workforce shifted into the machine section, which comprised 48.4 per cent of the production workforce by 1987. Most of the employment decline in the die-making section occurred among the class "A" die makers before the introduction of the WEDMs. The number of class "B" and "C" die makers and tool try-out workers remained relatively stable between 1977 and 1987. In the machine section, employment in many of the conventional machine-tool machinist job classifications had declined slightly by 1987, as discussed above. The increase in the number of production workers employed in the machine section resulted almost entirely from the addition of the WEDM operators, beginning in 1984.

The introduction of numerical control, notwithstanding the concomitant decline in the production workforce, was associated with a stabilisation of the remaining less skilled workforce and an increasingly dynamic internal labour market at Durable: inter-firm job mobility declined and intra-firm job mobility increased along with the introduction of the WEDMs during the post-recession recovery period.

Intra-firm job mobility increased at Durable during the post-recession recovery, the rates of both promotion and demotion tending to exceed those of the 1977-80 pre-recession era. (Transfers between job classifications with the same wage rates were virtually nil.) Promotion rates may have increased in response to the decline in the base of the percentage (that is, production employment) and the growing work volume that accompanied the introduction of the WEDMs. Similarly, the increase in the demotion rate may have resulted from a transfer of work from conventional machine tools to the WEDMs and from the reduction in "hand-fitting" by the die maker.

Turning to inter-firm job mobility, the rates of voluntary resignation at Durable seem to be lower than in the 1977-80 pre-recession era. Whether this decline resulted from intra-firm forces such as employee perceptions of increasing advancement opportunities stemming from growing promotion rates at Durable, and/or from such extra-firm forces as a deterioration in employment opportunities in the local external labour market, is unclear. None the less, the introduction of WEDMs seems not to have generated a sufficiently high level of worker alienation to have raised the resignation rate. With the exception of the relatively high level of lay-offs during the recession, the rates of lay-off and other attrition remained low and unassociated with the introduction of numerical control.

The impact of numerical control on the rate of accession (hiring new employees or recalling former ones) at Durable is unclear. On the one hand, the 1984-87 annual average accession rate of 12 per cent exceeded the 10.3 per cent annual average accession rate of the 1977-81 period, suggesting that the introduction of the WEDMs was accompanied by an increase in the accession rate. On the other hand, after the dramatic increase in the accession rate in 1984, immediately after the recession, the accession rate declined steadily, implying that the introduction of the WEDMs curtailed employment opportunities at Durable. The latter interpretation is consistent with the decline in production worker employment and the increase in worker productivity that were associated with the introduction of numerical control.

The introduction of numerical control had no impact on non-bargaining unit employment. The number of non-union employees in management and the engineering and accounting/personnel departments remained stable, along with the organisational structure.

2.6 Evaluation

The employment changes at Durable were accomplished with little involuntary displacement of workers, and little labour-management conflict. Several factors may have been at work here. First, the small size of the firm seems to have been conducive to the development of an historically personal and trustful relationship between labour and management. Consequently, union and management respected each other's needs. Workers did not resist the introduction of numerical control and management attempted to minimise worker displacement. Moreover, management tended not to lay off workers in any case because of the general shortage of skilled workers in the open labour market. Also, management trained existing workers for the new jobs, and accepted workers' self-motivated efforts to train themselves for programming and operating the NC equipment.

Second, the workers continued to be supportive of their union and, through collective bargaining with management, maintained a stable set of labour allocation and income security arrangements in the union agreement. In fact, neither labour nor management attempted to change these arrangements, which were intact before the introduction of numerical control. Workforce reductions were accomplished through an orderly and predictable procedure that was specified in the contract and designed to minimise displacement for any reason in favour of job reassignments and recall, whenever possible.

Third, technological displacement may have been minimised by the timing of the technological change in relation to the business cycle. Numerical control was introduced at Durable during a post-recession recovery, when the growing volume of work may have offset some of the employment decline that might have derived from the technological change. Also, career advancement opportunities improved during the post-recession recovery, albeit in an occupational structure with a shrinking proportion of high-wage jobs.

Although little forced technological displacement occurred at Durable, the union reaction to the introduction of numerical control was ambivalent. The workers felt that numerical control would improve Durable's market competitiveness and, therefore, their job security. Yet, they were apprehensive about the technological change and feared technological displacement. This ambivalence may have resulted from several factors. The workers' fears may have resulted from their lack of familiarity with and lack of participation in the selection and acquisition of the new technology. Greater participation of workers in these decisions might have allayed some of their fears. Moreover, the Durable workers, most of whom were IAM members, were unfamiliar with the model contract language on technological change designed by the national IAM to help workers increase their participation in decision-making on technological change. However, workers' fears may have been partly allayed by their trustful relationship with management, who informally reassured them that little displacement would result from the technological change.

In conclusion, the stable labour relations arrangements in the union agreement at Durable, arrangements that are commonly found in the American tool and die industry, led to minimal levels of involuntary technological displacement. These arrangements, coupled with the introduction of numerical control during a post-recession recovery and the traditionally personal and trustful labour-management relationship in this small family-owned firm, may have facilitated workers' acceptance of technological change and prevented disruption to their lives.

Notes

[1] Arne Kalleberg et al.: "The eclipse of craft: The changing face of labor in the newspaper industry", in Daniel B. Cornfield (ed.): *Workers, managers and technological change: Emerging patterns of labor relations* (New York, Plenum, 1987), p. 51.

[2] Christopher Ezell: *Typesetting: A sociological analysis of changing newspaper technology*, unpublished doctoral dissertation, Department of Sociology, Vanderbilt University, Nashville, Tennessee, 1987, pp. 103, 144.

[3] US Bureau of the Census: *US Census of Population: 1960*, Subject Reports, *Occupation by Industry*, Final Report PC(2)-7C (Washington, DC, Government Printing Office, 1963), pp. 57-61; *Census of Population: 1970*, Subject Reports, *Occupation by Industry*, Final Report PC(2)-7C (Washington, DC, Government Printing Office, 1972), pp. 329-36; *1980 Census of Population*, Subject Reports, *Occupation by Industry*, PC80-2-7C (Washington, DC, Government Printing Office, 1984), pp. 345-54.

[4] US Bureau of Labor Statistics: *Outlook for technology and manpower in printing and publishing*, Bulletin No. 1774 (Washington, DC, Government Printing Office, 1973), p. 7.

[5] ibid.: US Bureau of Labor Statistics: *The impact of technology on labor in five industries*, Bulletin No. 2137, microfiche (Washington, DC, Government Printing Office, 1982), p. 4; Andrew Zimbalist: "Technology and the labor process in the printing industry", in idem (ed.): *Case studies on the labor process* (New York, Monthly Review Press, 1979), pp. 105-12; Daniel Scott: *Technology and union survival: A study of the printing industry* (New York, Praeger, 1987).

[6] Michael Wallace and Arne Kalleberg: "Industrial transformation and the decline of craft: The decomposition of skill in the printing industry, 1931-1978", in *American Sociological Review*, 47 (June 1982), p. 317.

[7] US Employment and Training Administration: *Dictionary of Occupational Titles* (Washington, DC, Government Printing Office, 4th ed., 1977), pp. xvii-xviii, and *Selected characteristics of occupations defined in the Dictionary of Occupational Titles* (Washington, DC, Government Printing Office, 1981) pp. 469-73.

[8] See note 3.

[9] Patricia Roos: "Hot-metal to electronic composition: Gender, technology, and social change", in Barbara Reskin and Patricia Roos (eds.): *Gendered work and occupational change* (Philadelphia, Temple University Press, 1989).

[10] The source of data on the *Chicago Tribune* strike are 11 personal and telephone interviews which I conducted with three official representatives of Chicago Typographical Union No. 16.

[11] James Dertouzos and Timothy Quinn: *Bargaining responses to the technology revolution: The case of the newspaper industry* (Santa Monica, California, Rand, 1985), pp. 7, 27.

[12] Seymour Martin Lipset et al.: *Union democracy* (New York, Free Press, 1956); Harry Kelber and Carl Schlesinger: *Union printers and controlled automation* (New York, Free Press, 1967); Theresa Rogers and Nathalie Friedman: *Printers face automation* (Lexington, Massachusetts, D.C. Heath, 1980); Robert Jackson: *The formation of craft labor markets* (Orlando, Florida Academic Press, 1984); Roos, op. cit.; Ezell, op. cit.

[13] Kelber and Schlesinger, op. cit., pp. 68, 267-9.

[14] Kalleberg et al., op. cit.; Roos, op. cit.; Ezell, op. cit.; Tony Griffin: "Technological change and craft control in the newspaper industry: An international comparison", in *Cambridge Journal of Economics*, 8 (Mar. 1984), pp. 41-61; Michael Wallace: "Responding to technological change in the newspaper industry: A comparison of the United States, Great Britain and the Federal Republic of Germany", *Proceedings of the Thirty-Seventh Annual Meeting of the Industrial Relations Research Association*, Dallas, 28-30 Dec. 1984, pp. 325-32.

[15] *Proceedings of the Fifteenth Constitutional Convention of the AFL-CIO*, Hollywood, Florida, 3-6 Oct. 1983, Vol. II, *Report of the Executive Council*, p. 47.

[16] Wallace, op. cit.

[17] ibid.; Ezell, op. cit.; Roos, op. cit.; Kalleberg et al., op. cit.; Scott, op. cit., pp. 86-91.

[18] Kelber and Schlesinger, op. cit., pp. 37-47.

[19] 1949-51 contract between Chicago Typographical Union No. 16 and Chicago Newspaper Publishers' Association, pp. 2, 12.

[20] Daniel Cornfield: "Workers, managers and technological change", in idem: *Workers, managers and technological change*, op. cit.

[21] For national trends in unionised newspaper typographical worker wages, see US Bureau of Labor Statistics: *Union wages and benefits: Printing trades September 2, 1980*, Bulletin No. 2125 (Washington, DC, Government Printing Office, 1982), p. 6.

[22] *American Machinist*: "The 12th American Machinist Inventory of Metalworking Equipment 1976-78", Dec. 1978, pp. 133-7.

[23] *American Machinist*: "The 13th American Machinist Inventory of Metalworking Equipment 1983", Nov. 1983, pp. 113-18.

[24] David Noble: *Forces of production: A social history of industrial automation* (New York, Knopf, 1984); US Bureau of Labor Statistics: *Outlook for numerical control of machine tools*, Bulletin No. 1437 (Washington, DC, Government Printing Office, 1965), pp. 9-21; National Commission on Technology, Automation and Economic Progress: *The outlook for technological change and employment*, App. Vol. I of *Technology and the American economy*, Report of the Commission (Washington, DC, Government Printing Office, 1966), pp. 312-26.

[25] *American Machinist*: "12th Inventory", p. 136, and "13th Inventory", pp. 113-19; Harley Shaiken: *Work transformed: Automation and labor in the computer age* (Lexington, Massachusetts, Lexington Books, 1986), pp. 74-5.

[26] *American Machinist*: "12th Inventory", p. 136, and "13th Inventory", p. 134.

[27] *American Machinist*: "12th Inventory", p. 136, and "13th Inventory", pp. 113 and 118.

[28] US Bureau of Labor Statistics: Bulletin No. 1437, p. 26, and *Technology and labor in four industries*, Bulletin No. 2104 (Washington, DC, Government Printing Office, 1982), p. 21; US Office of Technology Assessment: *Computerized manufacturing automation: Employment, education and the workplace* (Washington, DC, Government Printing Office, 1984), pp. 59-60, 114, 281; National Commission on Technology, Automation and Economic Progress: *Outlook for technological change*, pp. 319-20; *Wall Street Journal*, 15 Feb. 1983, p. 44.

[29] US Bureau of Labor Statistics: Bulletin No. 2104, p. 27.

[30] US Bureau of Labor Statistics: *Productivity measures for selected industries, 1958-85*, Bulletin No. 2277 (Washington, DC, Government Printing Office, 1987), pp. 204-5.

[31] US Bureau of Labor Statistics: Bulletin No. 1437, p. 29, and Bulletin No. 2104, p. 21; National Commission on Technology, Automation and Economic Progress: *Outlook for technological change*, p. 307.

[32] US Bureau of Labor Statistics: Bulletin No. 1437, p. 47; US Office of Technology Assessment: *Computerized manufacturing automation*, pp. 5-6.

[33] US Bureau of Labor Statistics: *Employment and Earnings, United States, 1909-78*, Bulletin No. 1312-11 (Washington, DC, Government Printing Office, 1979), pp. 51, 244-53; *Supplement to Employment and Earnings* (Washington, DC, Government Printing Office, 1980), pp. 16, 78-81; *Supplement to Employment and Earnings* (Washington, DC, Government Printing Office, 1983), pp. 18, 78-81: *Supplement to Employment and Earnings* (Washington, DC, Government Printing Office, 1987), pp. 15, 58, 59.

[34] US International Trade Administration: *1988 US Industrial Outlook* (Washington, DC, Government Printing Office, 1988), Ch. 23, pp. 3, 7, 8, and *1987 US Industrial Outlook* (Washington, DC, Government Printing Office, 1987), Ch. 21, p. 10.

[35] US Bureau of the Census: *Annual Survey of Manufactures: 1966* (Washington, DC, Government Printing Office, 1969), p. 77; *1982 Census of Manufactures*, MC82-1-35C (Washington, DC, Government Printing Office, 1985), p. 7; *1983 Annual Survey of Manufactures*, M83(AS)-1 (Washington, DC, Government Printing Office, 1985), p. 20; *1985 Annual Survey of Manufactures*, M85(AS)-1 (Washington, DC, Government Printing Office, 1987), pp. 4, 20-1.

[36] *American Machinist*: "Builders attack machine-tool imports", Aug. 1982, pp. 45, 47; "Coalition targets downstream dumping", Feb. 1983, p. 23; and "Import quotas sought for machine tools", Apr. 1983, pp. 27, 29; US International Trade Administration: *1987 US Industrial Outlook*, Ch. 21, p. 2, and *1988 US Industrial Outlook*, Ch. 23, p. 3.

[37] *American Machinist*: "Filling the need for skilled workers", Special Report 712, June 1979, pp. 131-46.

[38] US Bureau of Labor Statistics: Bulletin No. 2104, pp. 27, 29.

[39] Daniel Cornfield: "Plant shutdowns and union decline: The United Furniture Workers of America, 1963-1981", in *Work and Occupations*, 14 (Aug. 1987), pp. 434-51; *Wall Street Journal*, 17 Aug. 1987, p. 8.

[40] US Bureau of the Census: *County Business Patterns*, United States summaries of the 1965-85 issues (Washington, DC, Government Printing Office, various years).

[41] Shaiken, op. cit., pp. 98-126; US Office of Technology Assessment: *Computerized manufacturing automation*, pp. 185-6; US Bureau of Labor Statistics: Bulletin No. 1437, p. 53.

[42] US Bureau of Labor Statistics: Bulletin No. 1437, pp. 36-7, and Bulletin No. 2104, p. 31; Shaiken, op. cit., pp. 98-126; US Office of Technology Assessment: *Computerized manufacturing automation*, p. 185.

[43] US Office of Technology Assessment: *Computerized manufacturing automation*, pp. 185-6; Shaiken, op. cit., p. 107.

[44] US Bureau of the Census: *Census of Population: 1970*, pp. 281-2, and *1980 Census of Population*, pp. 395-7; US Bureau of Labor Statistics: Bulletin No. 2104, p. 32.

[45] US Office of Technology Assessment: *Computerized manufacturing automation*, p. 144.

[46] US Bureau of Labor Statistics: Bulletin No. 1437, No. 50, and Bulletin No. 2104, pp. 29, 31; National Commission on Technology, Automation and Economic Progress: *Outlook for technological change*, pp. 161, 168.

[47] US Office of Technology Assessment: *Computerized manufacturing automation*, p. 136.

[48] Shaiken, op. cit., p. 67. See also US Office of Technology Assessment: *Computerized manufacturing automation*, pp. 194-5; National Commission on Technology, Automation and Economic Progress: *Outlook for technological change*, p. 168; US Bureau of Labor Statistics: Bulletin No. 1437, pp. 37-8, and Bulletin No. 2104, p. 31.

[49] Noble, op. cit., pp. 144-92.

[50] See e.g. US Bureau of Labor Statistics: *Industry Wage Survey: Machinery Manufacturing, February 1973*, Bulletin No. 1859 (Washington, DC, Government Printing Office, 1975), and *Industry Wage*

Survey: Machinery Manufacturing, November 1983, Bulletin No. 2229, microfiche (Washington, DC, Government Printing Office, 1985).

[51] ibid., table 41, pp. 75-6.

[52] US Bureau of Labor Statistics: *Industry Wage Survey: Machinery Manufacturing, February 1973*, table 3, pp. 16-17, and *Industry Wage Survey: Machinery Manufacturing, November 1983*, tables 12, 13, 15 and 18, pp. 24, 25, 28 and 32.

[53] *Wall Street Journal*, 27 Jan. 1983, p. 29, 17 May 1984, p. 33, and 4 Sep. 1984, pp. 1, 25.

[54] US International Trade Administration: *1988 US Industrial Outlook*, Ch. 23, p. 3.

[55] *American Machinist*: "13th Inventory", pp. 122-3; *Wall Street Journal*, 27 Jan. 1983, p. 29, 17 May 1984, p. 33, 4 Sep. 1984, p. 25, and 14 Sep. 1987, p. 1.

[56] IAM Research Department: *Report of the 1986 IAM National Machine and Manufacturing, Shipbuilding and Ship Repair Conference* (Washington, DC, Sep. 1986), pp. 8-9.

[57] Edward Kokkelenberg and Donna Sockell: "Union membership in the United States, 1973-1981", in *Industrial and Labor Relations Review*, 38 (July 1985), p. 524.

[58] IAM membership data are from the IAM Research Department, Washington, DC; Leo Troy and Neil Sheflin: *Union Sourcebook* (West Orange, New Jersey, Industrial Relations Data and Information Services, 1st ed., 1985), App. B, p. 10; Bureau of National Affairs: *Directory of US Labor Organisations*, 1986-87 edn. (Washington, DC, Bureau of National Affairs, 1986), p. 62. For the source of production worker employment data, see note 12.

[59] International Association of Machinists and Aerospace Workers: *Let's rebuild America* (Washington, DC, IAM, 1984), pp. 1-46, 141-51; Cornfield: "Workers, managers and technological change", op. cit., p. 12.

[60] International Association of Machinists and Aerospace Workers, op. cit., pp. 202-4.

[61] See Cornfield: "Workers, managers and technological change", pp. 8-24.

[62] IAM Research Department: *Report of the 1981 IAM Electronics and New Technology Conference* (Washington, DC, Apr. 1981), pp. 11-22.

[63] US Bureau of Labor Statistics: Bulletin No. 2104, p. 32; Daniel Cornfield: "Ethnic inequality of layoff chances: The impact of unionisation on layoff procedure", in Raymond Lee (ed.): *Redundancy, layoffs and plant closures: Their character, causes and consequences* (London, Croom Helm, 1987), pp. 116-40.

[64] Cornfield: "Workers, managers and technological change", op. cit., pp. 8-24.

[65] IAM Research Department: *Analysis of six (6) IAM area-wide "master" agreements in the machinery and manufacturing industry* (Washington, DC, Aug. 1986), sect. II, pp. 6, 13, 15, and *Analysis of selected wage and fringe benefits in IAM agreements with companies operating in the tool and die industry* (Washington, DC, Oct. 1988), sect. II, pp. 17, 19, 20, 21.

[66] IAM Research Department: *Analysis of six "master" agreements*, op. cit., sect. II, p. 16, and *Analysis of selected wage and fringe benefits*, op. cit., sect. II, p. 21.

[67] The source of data for this research on the Durable Dies Company consisted of company documents which were supplied by the company and a total of 11 personal interviews conducted with members of management, union die makers and machinists, and union officials.

[68] See the 1955-80 issues of US Bureau of Labor Statistics: *Analysis of work stoppages* (Washington, DC, Government Printing Office, various years).